Do Sparrows Like Bach?

THE STRANGE AND WONDERFUL THINGS THAT ARE DISCOVERED WHEN SCIENTISTS BREAK FREE

NewScientist

PEGASUS BOOKS
NEW YORK

DO SPARROWS LIKE BACH?
Pegasus Books LLC
80 Broad Street, 5th Floor
New York, NY 10004

First Pegasus Books edition 2010

Library of Congress Cataloging-in-Publication Data is available.

ISBN: 978-1-60598-114-7

10 9 8 7 6 5 4 3 2 1

Printed in the United States of America
Distributed by W. W. Norton & Company, Inc.
www.pegasusbooks.us

Contents

Introduction

Charles Darwin turned science (and religion) upside down with the publication of his research, but he has no chance of appearing in this book. In 2009, while the scientific establishment celebrated 200 years since his birth, we took a less enthusiastic view of his work. His theory of evolution may have startled the world and he may have been a meticulous experimenter but, as far as we can see, all that Darwin did to reach his conclusions was merely observe, record and rigorously document, before piecing together the – admittedly very clever – theory for which he is famous.

And that, quite frankly, doesn't put him in the same bracket as the rather more intrepid researchers you will find here – people like August Hildebrandt or Mark Grabiner.

In order to find a strong anaesthetic that worked without knocking a human out, Hildebrandt's tutor injected the student's spine with cocaine. Hildebrandt asked his tutor to tickle his feet – he felt nothing. He then handed his tutor a needle and asked him to stab him in the leg. There was no pain. So Hildebrandt demanded that a knife should be plunged into his thigh. When he still smiled benignly, the tutor stubbed out a cigar on his leg. Nothing. But Hildebrandt wasn't finished. To check the anaesthetic was remaining local to his lower body, his pubic and nipple hair was ripped out. Only the nipples hurt. After surviving hammer blows to his shins and a hearty tug at his testicles, presumably with some relief, he deemed the experiment a success. An important experiment for medical science, indeed, but a pretty bonkers way of going about it.

All of which goes to show, as surely everyone would agree, that Darwin had it easy. If he'd wanted to find fame in this book he should at least have injected himself with the droppings of one of his famous finches just to see if it caused him to grow an oddly shaped, but strangely useful beak. Even better, he should have dressed himself up as a finch, sat among them and waited to be fed.

By the same token Mark Grabiner is in the book, but Isaac Newton isn't. From observing a falling apple Newton came to the conclusion that something was pulling it down. But Mark Grabiner wasn't prepared just to sit and watch apples. He used gravity to more spectacular effect – tripping people up to see what kind of injuries they were likely to sustain and how such falls might be avoided. Coalface research at its best. And so Grabiner made it, Newton didn't.

Of course, August Hildebrandt's outlandish experiment had a positive and beneficial outcome. But he remains one of the few true success stories in this book. Plenty of science is dead-end research. We can learn as much from the experiments that fail as from the ones that succeed. Quite often research is dull and repetitive (even Darwin would have had days when he couldn't wait to break for a cup of tea). And anybody who has studied science at school will know what a serious and sometimes dull business it can be. Note: can be ... But occasionally, and sometimes unexpectedly, the routine becomes outrageous and outlandish, absurd even. That's why we are prepared here to celebrate the likes of Oscar Pike of Utah who proved that eating old odds and ends of food that people have found lying around in their cellars or outhouses needn't kill you. This has produced no discernible benefit to society but, as we said, nobody died (surprisingly) and it's earned him a place in this book.

The experiments of Hildebrandt, Grabiner and Pike are, in fact, a microcosm of what this book represents. Science can be intense, creative, often amusing, but beyond all that, science can fire the imagination like nothing else. And sometimes it's daft. That's because science researchers question everything and are prepared to tackle anything (just check out the chapter on love and sex). Scientists are used by governments to wage war, they are used by private companies to invent frivolous nonsense, and they are used by sports bodies to make better equipment and – in some cases – better athletes. How do you fancy airbags up your rectum to make you swim better? There is very little on this planet and beyond that remains untouched or unconsidered by the minds, if not the hands, of scientists.

So we've plundered the archives of 53 years of *New Scientist* magazine to bring you the science of farts and how to recycle your urine. You'll find out how to get rid of underpants in space and how two lager bottles brought a particle accelerator to a stuttering halt. And while we realise that science in many of the fields covered in this book will have moved on somewhat since most of these reports appeared, we are happy to print them almost as they first appeared in the pages of *New Scientist*. We are certain readers will spot that events and more recent research have rendered one or two a little obsolete – nuclear-powered aeroplanes anyone? However, we make no apology. Our bunch of free-thinking, no-holds-barred scientists will remain stuck in their own era, beacons to their limitless enthusiasm and seemingly boundless flights of fancy.

Of course, Darwin's great idea runs and runs and is developing still, something that you can't say for the majority of researchers and projects that appear in this book. Which is why, frankly, they deserve a second airing. The rigorous examination of the universe around us knows no bounds and is limited only by imagination, a surfeit of which imbues this collection of researchers.

Welcome to *Do Sparrows Like Bach?* which, if we'd had more room on the cover, might have been titled 'What scientists will get up to if you give them half a chance'.

Mick O'Hare

London, 2009

Especial thanks are due to Jeremy Webb, Helen Thomson, Catherine Brahic, Sandrine Ceurstemont, Liz Else, Richard Webb, John Hoyland, Paul Marks, Justin Mullins, Lucy Dodwell, Eleanor Harris, Ben Longstaff, Judith Hurrell, David Prichard, the staff of *New Scientist* down the years, illustrator Christian Northeast, and Paul Forty, Valentina Zanca, Andrew Franklin and Lisa Owens among many at Profile Books whose enthusiasm for this project has seen it overcome many obstacles. And our heartfelt thanks to all those contributors of the original articles featured in these pages. Lastly, special thanks must go to Sally and Thomas for putting up with the grumpy man by the computer ...

1 Mad inventions, mad ideas

It may be apocryphal, but it's a story too delightful to ignore. A minor researcher in the physics department attending a Princeton University party was writing in his notebook, and failed to recognise Albert Einstein, who had sat down next to him. 'What are you writing?' asked the great physicist. 'Whenever I have a good idea, I write it down to make sure I don't forget it,' replied the researcher. 'Perhaps you'd like to try it.'

Einstein shook his head sadly, and said: 'I doubt it. I have only had two or three good ideas in my life.'

Of course, not everybody who has an idea turns out to be an Albert Einstein, as this chapter reveals. Nonetheless, as Newton's third law of motion states, 'to every action there is an equal and opposite reaction'. Or, in other words, Einstein's genius needed its counterbalances, which – as you can find out below – the likes of Louis Douglas III seemed more than happy to provide. Douglas cleverly invented an entertainment aid for anybody using a public convenience. Eat your heart out Albert.

Of course, Douglas's urinary invention is scientifically about as far from Einstein's special and general theories of relativity as it is possible to get. But while relativity may have transformed our knowledge of physics, space-time and the cosmos, it never had a chance of making it into this chapter. That's because, in the main, the ideas here are more earth-grounded … and more ludicrous, outlandish and in the spirit of this book, entirely bonkers.

You see, Einstein cleverly thought about light bending around galaxies, while Sean McKee, instead, thought about making ice lollies in the shape of pop star Madonna and other sex icons to encourage people to lick them. Einstein considered how time could slow down when humans travelled at high velocities. Eugene Politzer, on the other hand, decided to use a laser to burn off his beard.

So none of them were ever likely to steal Albert's cleverness crown, but without them and their esoteric endeavours we wouldn't have this chapter, which pays homage to the archetypal idea of the 'mad scientist', beavering away in the lab, attempting to out-invent Heath Robinson.

The lonely, impoverished artist in his garret has his scientific counterpart in the undiscovered, bespectacled boffin. Hidden behind bubbling test tubes and clouds of noxious gas, he feverishly devotes a lifetime to coming up with a pointless idea that will baffle or, better still, outrage the world at large (not that he ever actually participates in the world at large). New Scientist *has, down the years, been keen to counter this stereotype – indeed you'll never find the word 'boffin' in the pages of the magazine. But, we have to admit, there are times when it would serve us well. Take this batch of ideas. So many things you really didn't know you needed …*

✳ Sobering thoughts

Normally it takes between 4 and 10 hours to sober up a drunk. In 1997, the University of Georgia tried patenting a high-speed method. The drawback? A catheter had to be pushed up the nose or the rectum.

The patient was given an enzyme to drink, yeast alcohol dehydrogenase, mixed with an acetate buffer to stabilise the stomach's pH. A catheter was then inserted. This fed pure oxygen into the small intestine. The enzyme and oxygen together would accelerate the conversion of blood alcohol into acetates and regenerate enzymes in the liver. This technique could lower the concentration of alcohol in the blood to below the lethal level of about 1 per cent in less than half an hour.

Clearly there are numerous advantages to be gained from this form of internal ventilation ...

✲ Sporting chance

In his book *Gut Reactions: Understanding symptoms of the digestive tract*, W. G. Thomson wrote that in preparation for the 1976 Olympics, East German swimmers had 1.8 litres of air pumped into their colons to improve buoyancy. Thompson said: 'It apparently helped crawl and backstroke specialists, but a breaststroker complained that his gas-filled gut caused his feet to stick out of the water. Perhaps sports authorities will need to test athletes for flatus, as well as steroids.'

And presumably their tummies suffered somewhat from bloating ...

✲ Gutted

In 1999, James Stage of Aberdeen came up with a simple idea for helping overweight people hold their bellies in. A small device that looked like a pager was worn on the trouser belt. The 'pager' had a microswitch which rested against the stomach. Pressure on the switch closed a circuit and current flowed through a small motor with a deliberately unbalanced weight on its spindle. So every time the wearer let their belly sag, the motor caused uncomfortable tickling around the waist. This encouraged the wearer to hold their stomach in.

Some inventors are prepared to plumb even greater depths.

✖ Aim 'n' squirt

Back in 1990, Louis Douglas III of San Francisco filed a patent application for what he tastefully described as 'an amusement device for urinals'.

The problem with urinals, said the inventor, especially in nightclubs, was that people who used them were often inebriated and hence their aim was poor. What was needed was something to capture the attention and imagination of even the most drunken user. The patented solution was a pressure and heat sensor built into the base of the bowl, to sense the presence of warm urine. This then triggered an electrical circuit, which produced an audible or visible signal.

In its simplest form, this could be a lamp or a buzzer. But more impressive systems would have an array of lights and a sound synthesiser. 'This,' said the inventor, 'induces the user to express his or her artistic talents by creating an appropriate light show or symphony through a loudspeaker.'

And the following garment might come in handy when fitting the new system.

✻ Shirt rolls up its own sleeves

In 2001, a shirt which rolled up its own sleeves when the wearer got too warm was unveiled by a tech-savvy Italian fashion house. And what's more, its inventors claimed it would never need ironing.

The fabric for the prototype shirt was woven from fibres of the shape-memory alloy nitinol, interspersed with nylon. The alloy could be deformed, and then returned to its original shape when heated to a certain temperature. It was this shape memory property that was key to how the 'memory metal shirt' worked. 'The sleeve fabric is programmed to shorten as soon as the room temperature becomes a few degrees hotter,' said Susan Clowes, a spokeswoman for Corpo Nove of Florence, the shirt's developer. 'Even if the fabric is screwed up into a ball, pleated and creased, a blast from a hairdryer pops it back to its former shape,' Clowes said. This meant the shirt could be 'ironed' as you wore it. 'It's a traveller's dream,' she said.

But fashion victims could not readily expect to buy one of Corpo Nove's intelligent shirts on their next shopping trip. The prototype cost around £2,500 to make, and was available in any colour you liked – provided you had a tendency to wear metallic grey. 'But it does look distinctly bronze-coloured in some lights,' said Clowes.

Some inventions, of course, are less easy to categorise. Although one might consider placing this inventor in the file marked 'logically challenged'.

✱ The ears have it

Some animals prick up their ears to get messages across. Karola Baumann of Düsseldorf believed, therefore, that we could communicate better with animals if they could see us pricking up our ears. In 1998 Baumann and colleagues described a device that they hoped would transform the wearer into a latter-day Doctor Doolittle. It was a skullcap with two short 'masts' at either side over the ears, each carrying a large replica of an animal's ear. These ears could be moved as the human 'talked' to an animal, so the beast's attention was, Baumann claimed, grabbed and held.

And is there really a social requirement for this next idea?

✱ Here we go round the dining table

Want to get away from a crashing bore at Christmas dinner? Or escape from an unruly toddler who's flicking food at you? In 2001, Paolo Rais came up with a solution: a dining table whose chairs kept moving so that nobody spent more than ten minutes sitting opposite anyone else.

It was while sitting at a wedding dinner that Rais, a civil engineer from Lugano, Switzerland, realised that the traditional rectangular dining table meant he could only talk to a few friends seated around him. 'So I wondered how I could help people talk to more of the people around a long table,' he said.

His answer was an 18-seater, mains-powered rectangular table that cleverly concealed what might otherwise be an ugly drive mechanism in a neat central pedestal. An electric motor drove two hidden chains: one hauled the chairs around while the other, beneath the wooden tabletop, pulled around wooden trays on which you placed your food or documents. 'So your tray of stuff always stays in front of you,' said Rais. The chairs' connection to the drive chain was covered by a foot platform, so there was no chance of getting your feet trapped in the mechanism. And at a steady speed of 9 centimetres per minute, diners didn't really feel they were moving at all, said Rais – and they could leave or join the table at any time without problems.

Rais tested his table out in a restaurant and a hotel meeting room – with mainly positive reactions. And he also had some grander designs. 'I've written to the British royal family, because such a table would be a great way for the Queen to meet all her guests at banquets,' he said. 'But I have had no reply.'

Nor, indeed, is there a pressing requirement for any of the following, all of which are indicative of the desire of scientists to fulfil social needs which we'd suggest don't even exist …

☀ Trials of life

In 1990, parents looking for a toy to educate their children in the facts of mammalian life needed look no further than the 'toy birthing apparatus with chugging-like delivery motion' patented by Douglas Raymond. His toy dog contained a complicated combination of pistons, springs, air-bleed orifices and grommet seals. When a spring was wound up and released, the pistons chugged backwards and forwards forcing a collection of fetal dolls, fashioned as puppies, down a tube and out through a spring-loaded valve at the animal's rear. Each delivery was accompanied by loud barking from the dog and yelps from the newborn pups.

⚛ Dogs of stun

In the 1990s, people were becoming so litigious that even criminals were suing if they were bitten by a guard dog. In 1997, things had reached such a state, according to Harvey Allen, William Buerke and Gary Erwin of Orange, California, that guard dogs had to be muzzled. However, as they so succinctly pointed out, 'criminal suspects are less likely to submit to apprehension by a dog that is unable to bite'.

So the inventors proposed using the dog as a mobile stun gun. Beneath the dog's muzzle was a leather or plastic pod containing a pair of metal electrodes. These were connected to a battery-powered circuit that generated brief, high-voltage pulses. When the dog's handler pressed a button, sparks jumped between the electrodes. The dog would prod the alleged criminal and shock the victim into submission.

⚛ Off the leash

Keeping pets can be expensive and inconvenient. Back in the 1990s, Daniel Klees and Terri Shepherd of Illinois believed they had the answer. They filed patents for a 'novelty leash' that let people pretend they had a pet.

The leash was a length of thick cord like an ordinary lead, but with a core of strong wire so it kept its shape. The lead contained batteries and a loudspeaker that was pre-programmed to make a variety of animal sounds, such as a dog barking or cat meowing. Owners could take walks by holding the leash out in front of them like a metal detector. The inventors told purchasers that they would need 'a degree of imagination'.

✳ Safety brolly

An ideal present for the anxious is an umbrella that protects them from fire. In 1991, Taiwanese inventors proposed such a brolly: it would look like an ordinary umbrella, but would be clad with ceramic insulation. So if the owner was caught in a fire, the brolly would protect the head. If that failed, said the inventors, the brolly could be used as a parachute 'to help the user escape from a high-rise building'.

✳ File nirvana

Do you feel uneasy when you delete files from your computer? Does it seem somehow wrong to consign all those once-valued words to a state of nonexistence, simply by pressing a button on your keyboard? In 1997, a Buddhist monk in Japan set up a virtual temple on the internet to have memorial services performed for obsolete software, failed business projects and information that had been lost or deleted.

Shokyu Ishiko, the chief priest of the Daioh Temple in Kyoto, dedicated his virtual Information Temple to Manjusri, the Buddhist incarnation of wisdom. He also offered counselling on his website.

We reckon the Buddhist approach is more socially acceptable than attacking your PC with your phone handset but we're not sure it's as satisfying. And will the Daioh Temple be able to cope with Vogon poetry?

✺ Poetry website goes from bad to verse

Vogons, fans of *The Hitchhiker's Guide to the Galaxy* will recall, wrote poetry so bad it could kill. In 2003, an experiment to create poems on the web looked set to automate the awfulness of Vogon verse.

David Rea of Greenwich, Connecticut, wrote a program that allowed a poem to evolve, to see if people with diverse tastes in poetry could work together to create attractive verse. Rea's program started off with 1000 'poems', each comprising four lines of five randomly chosen words. People visiting the website chose between two randomly selected verses from the population. The bad ones were killed off and the fittest – those with the most positive votes – underwent further evolution.

Each word within a verse was thought of as a poetic gene. There were a possible 30,000 words, and as people voted, some genomes proved more popular than others as they formed semi-meaningful phrases. So the fittest verses were 'mated' to form new verses, and the offspring again put to the public vote.

With more than 16,000 votes in, Rea believed poetic structure was emerging. But in evolutionary terms, the poems were still a metaphor short of a mudskipper. When this story first went to press in 2003, one poem read: 'You with life down the swords / How quieting tressed / Prince held by posers / Could be honking fight trekking.'

✺ Real ice

In their efforts to diversify into other technologies in the 1980s, Japanese steel company Nippon Kokan (NKK) came up with ice that crackled. NKK scientists brought back ice from the Antarctic for research but found that when used in alcoholic drinks, it crackled distinctively thanks to the release of air bubbles, which had been trapped inside the ice for thousands of years. NKK managed to improve on nature – their ice, called Exice and based on the Antarctic original, crackled even louder. A glass of whisky on the rocks at about 40 per cent alcohol could produce a 70-decibel crackle every second, while a cocktail with only 11 per cent alcohol gave out 65 decibels every 2 seconds.

✳ Burnt offering

In 1986, French inventor Eugene Politzer believed that shaving could, one day, be a matter of burning hair away, rather than cutting it.

He filed patents around the world on a laser shaver. He said that this would overcome the problem of 5 o'clock shadow, where new growth starts showing through even after a close conventional shave. According to Politzer's patent, the power for the laser would be generated by a fixed supply connected to a hand-held unit, which contained a small helium neon tube. This would beam laser light along the inside of a protective grid mesh, similar to that used in a conventional razor. Any hairs protruding through the mesh would conduct heat down to their base and be burnt off. To stop the mesh overheating and burning the owner, a small motor and fan would circulate air through the mesh.

Presumably Politzer was already a big fan of Jan Louw …

☀ Cutting edge

In the mid-1980s, a South African designer, Jan Louw, devised an attachment for a vacuum cleaner which let people cut their own hair. The cutter looked like a hair dryer but was connected to the hose of your vacuum cleaner. The air sucked in through the open end of the 'hair dryer' nozzle would drive a turbine which rotated one blade over another stationary blade to mimic the action of scissors. The user would slide the end of the nozzle over the scalp and hair would be sucked in to be sliced by the blades. The shorter the nozzle length, the shorter the haircut. Cut hair ended up in the bag of the vacuum cleaner.

Meanwhile, in a research institute far away from those trying to rectify society's most obvious ills, a whole different group of inventors was working on ways to help us get about, safely or otherwise.

☀ The saucer at Platform 9 ...

Anyone spurred on by an interest in flying saucers to think of building one for themselves could do little better than to turn to British patent number 1,310,990. Filed in 1970, this described an atomic-powered, saucer-shaped space vehicle. But the patentee, none other than national railways operator British Rail, didn't pay the renewal fees so the patent lapsed, meaning anyone would be allowed to construct the saucer according to the specifications. These explained that powerful electromagnets deflected charged particles produced by a thermonuclear fusion reaction to create lift and drive. Electrodes in the pulsed field were bombarded by charged particles and generated electric power for the craft. British Rail said the idea was less crazy than it sounded, insisting it was invented by an experienced nuclear physicist.

⚛ On the hoof

In 1981, Philip Barnes of Cambridgeshire, UK, filed a patent on what was literally a one-horsepower road vehicle. A small minibus with seats for five passengers and a driver was powered by a horse which walked along a conveyor belt in the central aisle of the coach. The conveyor was an endless, free-running loop with its axles connected to a gearbox and alternator. The horse, tethered by a harness, would plod along the conveyor belt to propel the vehicle and charge its battery, while the driver steered with a wheel.

The inventor claimed several advantages. The horse would be walking on a clean, flat surface so could not damage its hoofs. A thermometer under its collar would alert the driver to any overheating of the 'engine'. The horse would be fuelled from a food box under its head and panels would protect the passengers from stray kicks.

The vehicle would be started by prodding the horse with a mop.

⚛ How to brake blinking fast

Many car accidents occur because of the relatively long time drivers take to react when driving. Drivers view the road ahead, pass the picture from the retina to the cortex and the brain, then after a lag caused by fear or other psychological reasons, the brain computes the action to be taken and sends messages to the foot to apply the brakes. The total delay can be as much as half a second. Back in the 1960s, Professor

Vadovnik of Ljubljana University in Yugoslavia suggested using other muscles closer to the brain to initiate braking under emergency conditions. He elected to use the eyebrow muscles, because these are small and can react in 0.1 seconds if suitably trained. He mounted a pair of electrodes on a spectacle frame – these were kept in contact with the eyebrows by small springs. When drivers encountered an emergency situation, they were expected to blink rapidly to send an electrical signal to the car's braking system. Professor Vadovnik believed that practised blinkers would be able to cut braking time by 0.3 seconds.

✳ I'm not moving

In 1998, Samsung proposed a voice-controlled car with a difference. The dashboard used standard voice-recognition circuitry to respond to such phrases as 'start engine'. But a sample of the driver's speech, when sober, was also stored. When the driver spoke to the car, a comparator program looked for any sign of a slur. If the verdict was 'sober', the engine sprang into life. If 'drunk', a speech synthesiser warned: 'Please don't drink and drive as you are putting your life and property at risk.'

And, perhaps more importantly we reckon, the lives of pedestrians and other road users, even if they were cleaning their teeth with this ...

✳ Warning smile

In 1967, from the research laboratories of American industry there came a toothpaste which glowed in the dark and reflected the headlights of motor cars. It was the first tooth-paste that could be advertised as making a definite contribu-tion to road safety – so long as those who used it remembered to walk facing the traffic and to keep grinning.

Bulk transport

In 1994, a design firm called Teague, in Redmond, Washing-ton, proposed carrying tourist-class air passengers in cylin-drical pods. The pods would be loaded and unloaded from 'high-capacity' planes with the passengers inside. They would include self-contained entertainment systems, plus an 'adjustable form-fitting cushion'. A life-support system would provide air to the enclosed passenger.

Down the years a few ideas seem to have gone up in smoke, and not just figuratively.

Smoker's cough

In 1992, Wan Chung of Taiwan filed a patent application for an ashtray designed to turn against smokers. It had a slot at the side for a box of matches, and when a smoker picked up the box, the ashtray would make a coughing sound and warn against smoking.

⚛ Fag end

In 1986, Danish inventor Kaj Jensen believed that his self-extinguishing cigarette was a world-beating idea. He patented it in no fewer than 37 countries.

Discarded smouldering cigarettes are both unpleasant and dangerous, claimed Jensen. So his cigarette had a small ampoule of water buried inside it, just ahead of the filter tip. The ampoule was made from polythene and had a weakened thinner wall facing the burning end of the cigarette. As soon as the glow reached the ampoule, it melted, releasing the water and extinguishing the cigarette.

The ampoule could be placed further down the cigarette, making it impossible to smoke the last part of it and therefore, it was hoped at the time, making smoking safer.

And imagine having to shovel coal every time you wanted to watch TV.

⚛ Coal-fired television set makes its debut

In 1896, powdered coal was burnt in an electrolytic cell to generate electricity. Unfortunately, the efficiency was dismally low, about 1 per cent. But in 1965 the idea was given a new twist by scientists at the Westinghouse Research Laboratories in Pittsburgh, who developed a television set working from a fuel cell which consumed powdered coal. The experiments were aimed at developing a large-scale coal-burning fuel system.

Of course, in these carbon footprint-conscious times, coal is a bit of a no-no. But is it any less antisocial than tenderising your T-bone by blowing it up?

✳ Beef à la dynamite

A new way of tenderising meat devised in the year 2000 had just one snag – it involved a hefty explosion.

Most people tenderise meat with a culinary hammer, bashing it repeatedly to break down muscle fibres. Or you can add meat-tenderising powder, which contains an enzyme that digests muscle fibre and connective tissue. But how do you tenderise meat on an industrial scale?

Researchers at the US Agricultural Research Service in Beltsville, Maryland, thought they had an answer. They had been blasting meat with water at explosive pressures. And they found their process also killed food-poisoning bacteria, such as *E. coli*, in the meat. 'We think it is probably rupturing the bacterial cell walls,' said lead scientist Morse Solomon.

The process worked by sending a shock wave through the meat to bust the tough, chewy fibres. To create the shock wave, researchers placed a slab of meat on top of a steel plate at the bottom of a water-filled plastic garbage can. Then they detonated an explosive – equivalent to about a quarter of a stick of dynamite – inside the can. The water transmitted the shock wave through the meat, but the unfortunate garbage can was blown to smithereens.

Solomon said the shock waves penetrated the entire cut of meat, so bugs deep inside it were killed, achieving a thousand-fold reduction in bacteria levels during tests. The process worked best on small, garbage can-sized batches. A larger tank didn't work as well, and the meat had to be packaged in robust containers so it wasn't destroyed.

Randy Huffman of the American Meat Institute in Arlington, Virginia, welcomed the idea but said: 'The real challenge will be getting this implemented in a real-world solution.'

The culinary thread continues with a diet of sediment.

✳ Edible mud

In 1957, it was thought that mud dredged up from the bottom of Lake Victoria, 20 metres down, might be turned into food for pigs and chickens, according to the director of the East African Fisheries Research Organisation (EAFRO), Mr R. S. A. Beauchamp.

After studying the composition of the fauna and flora that are found in Lake Victoria, Beauchamp's colleagues discovered that there was not as much plankton as they had expected in the water, because of a shortage of sulphur. In lakes in temperate climates, millions of dead plankton, weeds and water animals fall to the bottom every year and when they decompose they return to the water the elements they took from it to grow.

But the mud found on the bed of Lake Victoria was very slow to decompose because of lack of oxygen. This meant that at the bottom of the lake, locked up in layers of mud which ran to tens of metres deep, was the accumulated richness of thousands of years' deposits. The water lying above was infertile for lack of nutrients.

EAFRO suggested that swamps could be filled up with the rich lake mud to make good land for market gardens, but Beauchamp also believed that, dried and powdered, the mud could become nutritious food for pigs and poultry because it contained numerous valuable minerals and also considerable amounts of protein.

The dried mud was, apparently, not unpleasant to eat – Beauchamp tried it himself and offered it to both his friends and family in order to prove his point.

While Mr Beauchamp displayed excellent scientific insight, his social skills left a little to be desired. And ironically, at the same time as he was attempting to unearth dead matter, an inventor in Winnipeg was doing the opposite.

✺ Funerals without fuss

Back in the 1950s, a gentleman in Winnipeg devised a method of shrinking an adult human corpse to the size of a baby. He claimed that this would be an excellent way to dispose of dead bodies because it would be possible for the shrunken corpses to be slid into underground tubes and buried without any of the fuss that goes with conventional burial and without taking up too much space.

His shrinking process depended on the application of high pressure to the corpse. This, he said, had the effect of squeezing all the fluids out of the body. The resultant mass of flesh could then be compressed to any desired shape inside a mould, similar to those used for moulding polythene.

All this was, of course, highly ingenious but it was uncertain why the inventor thought that his proposed form of burial was in any way better than the prevailing and very popular method of cremation. It might have been that he thought the new method would appeal to those who were worried by the high cost of buying plots in cemeteries and who did not wish their dead body to be incinerated.

It may also have appealed to those whose beliefs suggested that at some point in the future they could expect to be physically reincarnated. Whatever the benefits, however, it appeared that the inventor struggled to find a volunteer for his process.

It seems that the people of Winnipeg had a penchant for wild ideas in the 1950s. Another inventor in the city was also struggling to make headway with his invention. It consisted of a gigantic cake of soap 15 feet long and 3 feet wide. His idea was that you wet yourself with water before sitting on the soap and sliding up and down to wash yourself. As much fun as licking a lolly that looks like Madonna? Surely not?

✳ How to mould an X-rated lolly

In 1992 a new wave of 'adult' ice lollies promised to liven up summer thanks to a new way of moulding detailed shapes. It was hoped that jet-stream extrusion moulding could provide British lolly lickers with risqué lines including lifelike images of Madonna and other sex symbols.

Sean McKee, a researcher at the University of Strathclyde, said the technique could pave the way for companies to increase the size of the £150 million water ice lolly market by targeting adults. 'Just imagine Madonna or Marilyn Monroe on a stick,' he said. 'We could get at least half the adult population wanting to lick ice lollies.' Spaniards had already taken to licking Hot Lips, a red lolly shaped like a large pair of lips.

Since Snofrut, the first mass-market ice lolly, went on sale in 1922, lolly companies had struggled to improve their moulding techniques and had only been able to make blurred shapes. Production lines poured flavoured water into crude rubber moulds which were frozen and turned inside out to release the lolly.

Jet-stream extrusion moulding squirted a slush of almost-frozen water ice into a hard mould at high speed. How the lolly came out of the mould was a trade secret. But Birds Eye Wall's, a division of Unilever, put the world's first extrusion-moulded lolly on sale in the UK in 1992 with Boomy, a sort of fruit kebab that was distinctly unsexy.

McKee said one of the biggest production problems for extruded lollies was structural weaknesses caused by air bubbles trapped in the mould. 'If the character's nose is missing, there will be no repeat sale,' he said.

Clarke Foods, which makes Lyons Maid lollies, was dubious about the value of the adult market. David Brown, a marketing director, thought that lollipops were a family product. 'What sort of adult audience you would attract I'm not quite sure,' he said. 'We would not want to do anything that was too risqué.'

We second that, after questioning when the lolly likenesses of Brad Pitt and David Beckham were going to be produced for those who prefer licking males. Still, it seems novelty items have no boundaries …

✳ Mozart's bra

The bicentenary of Mozart's death in 1991 was celebrated by Japanese electronic wizardry. A lingerie firm in Tokyo made a musical bra containing a microcircuit in the clasp and a small loudspeaker which fitted into the wearer's armpit. Hooking the bra up switched its audio system on, and it played a 20-second blast of Mozart. This was part of a miniature *son et lumière* display, because the bra also featured tiny flashing lights that winked during the music.

Presumably you wouldn't want to be wearing the bra when trying out the following.

☀ Vital statistic

In 1981, two inventors in the US, Jack Grossman of California and Leonard Roudner of Chicago, claimed to have perfected a scientific system for measuring the size of female breasts with great accuracy. They were so proud of their results that they patented them.

It seems that, in the past, this vital statistic had usually been estimated by eye. The only scientific approach had been to immerse the breast in a container of water and to measure the volume of liquid displaced. According to the inventors, this was inaccurate because breasts tend to float.

The solution was a circular sheet of transparent flexible plastic with a single radial cut and a series of radial calibrations. The circular sheet was folded into a cone and placed over the breast. The cone was then tightened until there was all-round contact. The calibrations gave an instant readout of the breast's volume in millimetres.

Incidentally, breasts leave unique patterns on surfaces in much the same way as fingerprints. Even more surprisingly, it seems that jeans do too.

☀ In their jeans

In 1998, the FBI found an ingenious way to catch crooks – by looking at their jeans.

Scientists from the bureau reported that every pair of blue jeans has a unique wear pattern. The FBI actually used this 'bar code' to place a suspect at the scene of a crime.

Richard Vorder Bruegge, a forensic scientist at the FBI laboratory in Washington DC, and his colleagues developed the technique while helping to identify suspects who were robbing banks and setting off bombs in Spokane, Washington. In April 1996, one of the gang was caught on film. He was wearing a mask, but part of his trousers was visible.

When the photograph was enlarged, Vorder Bruegge noticed light and dark lines running across the seam of the man's jeans. His team found that the pattern originated from slight imperfections introduced when the trousers were made. Workers sew the seams by pushing the fabric through a machine, and the irregularity of that motion stretches and bunches the fabric. The dyed layer of cotton in the raised portion is worn away, creating white bands.

The patches are more striking on jeans than other types of trousers because they are often allowed to become extremely worn. 'People just keep wearing them,' said Vorder Bruegge.

The FBI analysed the jeans of suspects in the Spokane case. One pair had a pattern with over two dozen features that matched the jeans photographed by Vorder Bruegge's team. At the trial, the defence called in a used jeans exporter as an expert witness who claimed the patterns were common to all jeans. He showed the court 34 similar pairs, but in each case the FBI could distinguish them from the accused's. The suspect was convicted.

Some ideas, on the face of it, seem to make complete sense. It's only when the practicality of putting them into action rears its contrary head that you see them in a different light.

⚛ Round the bend

Back in 1965, a circular runway for airports was being considered by the US Navy. It was thought to have some potential advantages even if it would be more expensive than normal runways. One advantage was that it would save one-third of the space occupied by a conventional airport.

The idea was to match the circumference of the runway to the landing speeds of aircraft – for a large jet, a circumference of about 60,000 feet would be needed, allowing it to turn without toppling over. One more benefit was that if crosswinds were favourable, six aircraft could land simultaneously at different, equally spaced points on the runway.

⚛ A tube train that splits down the middle

Underground railways are the fastest way of transporting large numbers of passengers around cities and, in 1969, French engineers proposed a way of making the Metro even faster. The plan was for an underground train to run continuously, serviced by 'ferry' coaches from the stations to take passengers on and off.

The 'AT 2000' looked like an ordinary underground train split lengthwise down the middle. The left-hand section would be fitted with seats in the normal way, and would never stop at stations. The right-hand side would have no seating. Instead it would act as a loading and unloading platform and stop at every station.

The two sections would travel together like a railway coach with a corridor. Between stations, doors joining the two sections would open for a short period to let passengers move from one section to the other. Passengers wishing to get off at the next station would move into the 'corridor' section. Those who had just got on would take their seats in the non-stop section. As the vehicle approached the station, the 'corridor' section would separate and halt for the unloading and loading of passengers. The other section would continue and link up with another 'corridor' section already loaded with passengers.

❈ Artificial tornado plan to generate electricity

Most of us know that tornadoes are unpredictable, uncontrollable and dangerous. But a Canadian engineer believed they could be the future of electricity generation. He wanted to make electricity from artificial tornadoes.

In 2008, Louis Michaud, a retired petroleum engineer in Sarnia, Ontario, planned to use the waste heat from conventional power plants to create an 'atmospheric vortex engine' – a small, controlled tornado that would drive turbines and generate electricity. 'I'm confident that we could control these things,' he said. Michaud also thought solar-powered tornadoes generated using the sun's heat could work.

His design was a circular wall 200 metres across and 100 metres high without a roof. Air carrying the waste heat would be blown in from vents on the sides, spinning around the walls into a vortex that becomes just like a real tornado. Once started, the vortex would draw in more hot air from vents in the wall, pulling it past turbines and generating electricity.

Michaud calculated that a vortex engine of this size would create a tornado about 50 metres in diameter and generate between 50 and 500 MW of electricity. He patented the idea in 1975 and has continued to work on it ever since.

Is it such a crazy idea? If only Michaud had befriended Professor

Dessens, he might have been richer, faster. He'd also have a carbon footprint to rival Detroit's. However, the idea of creating potentially disastrous major weather phenomena was so like the script of a dodgy 1950s sci-fi B movie that we decided to name the book after this next concept.

✳ Cloud-making experiment reaps a whirlwind

In the early 1960s, a French scientist, J. Dessens of the Observatoire du Puy de Dôme, Clermont University, accidently discovered a way to make tornadoes artificially – and therefore a means of studying the conditions under which they arose.

On a plateau in the south of France the observatory built an apparatus which was originally intended for making artificial cumulus clouds. It was called the Meteotron and consisted of an array of 100 burners spaced over an area rather larger than a football field. Fuel was pumped into them and, together, they consumed about a ton of oil a minute, producing the very considerable power of some 700,000 kilowatts. In operation the device produced a thick column of black smoke that permitted observations of the resulting upward air currents.

During one experiment there appeared, besides the main mass of smoke, what seemed to be a black tube of whirling smoke. This tornado, about 30 or 40 feet across and up to 700 feet high, seemed to form about six minutes after the burners had been lit and subsequently moved away from the apparatus at the speed of the prevailing wind.

Later, the research team attempted to reproduce the tornado conditions with only half the burners operating, but under very unstable atmospheric conditions. After about a quarter of an hour's heating, they started a strong whirlwind in the centre of the apparatus about 130 feet across, with a bright tube a yard in diameter at its centre. It was so powerful that some of the burners were extinguished.

This is real mad boffin stuff of the kind you thought existed only in movies. Completely bonkers, just like the one that follows.

⚛ Human cannonballs

The old circus trick of firing a person from a cannon was being considered by the US Defence Advanced Research Projects Agency (DARPA) in 2006 as a way to get special forces, police officers and firefighters on to the roofs of tall buildings in a hurry.

A ramp with side rails would be placed on the ground near the target building at an angle of about 80 degrees. A (very brave) person would then sit in a chair, like a pilot's ejector seat, attached to the ramp. Compressed air from a cylinder underneath would be rapidly released to shoot the chair up the ramp's guide rails. At the top the chair would come to an instant halt, leaving the person to fly up and over the edge of the roof, to land (one hopes) safely on top of the building.

Of course, the trick would be to get the trajectory just right. But the DARPA patent suggested a computer could automatically devise the correct angle and speed of ascent. It also claimed that a 4-metre-tall launcher could put a man on the top of a 5-storey building in less than 2 seconds. Most of us would probably prefer the stairs.

As a visitor to www.newscentist.com later said, 'Aim really is important here'. Of course DARPA thrives on adversity of the kind thrown up on a plate by the ideology of the Cold War. That's right – if it wasn't boffins, it was Dr Strangelove. Mutually Assured Destruction provided plenty of opportunities for mad research, as does the whole field of weaponry and espionage generally.

☀ The shouting bomb

Back in 1957, the Americans armed themselves with a shouting bomb which could address large crowds as it dropped to the ground on a parachute. According to its designers, military tactics at the time required a means of communicating propaganda or instructions to groups of soldiers or civilians.

The bomb was said to be about 9 feet long and could be dropped from 60,000 feet. As it fell, it sprouted parachutes. At 4000 feet a tape recorder was automatically switched on and the bomb proceeded to deliver a 3-minute lecture. Trials showed that the bomb could be heard over an area greater than half a square mile.

☀ US military creates indestructible sandwich

First came the atom bomb, the stealth bomber and the airborne laser. Then in 2002 came one of the US military's most fearsome weapons: the indestructible sandwich.

Capable of surviving airdrops, rough handling and extreme climates, and just about anything except a GI's jaws, the pocket sandwich was designed to stay 'fresh' for up to three years at 26 °C (about the temperature of a warm summer's day), or for six months at 38 °C (just over body temperature).

For years the US army had wanted to supplement its standard battlefield rations, called 'Meal, Ready-to-Eat' (MRE), with something that could be eaten on the move. Although MREs already contained ingredients that could be made into sandwiches, these had to be pasteurised and stored in separate pouches, and the soldiers needed to make the sandwiches themselves.

'The water activities of the different sandwich components need to complement each other,' explained Michelle Richardson, project officer at the US Army Soldier Systems Center in Natick, Massachusetts. 'If the water activity of the meat is too high you might get soggy bread.'

To tackle the problem, researchers at Natick used fillings such as pepperoni and chicken, to which they added substances called humectants, which stop water leaking out. The humectants not only prevented water from the fillings soaking into the bread, but also limited the amount of moisture available for bacterial growth. The sandwiches were then sealed, without pasteurisation, in laminated plastic pouches that also included sachets of oxygen-scavenging chemicals. A lack of oxygen helped prevent the growth of yeast, mould and bacteria.

Soldiers who tried the pepperoni and barbecue-chicken pocket sandwiches found them 'acceptable'.

The mother of all firework displays

In 1998, Dave Caulkins hoped the millennium celebrations would go with a real bang, if he got his way.

Caulkins, a computer network manager based in Los Altos, California, devised a plan to use obsolete Inter-Continental Ballistic Missiles (ICBMs) to launch artificial meteors. They would create stunning pyrotechnic displays up to 20 kilometres across, he said.

Caulkins outlined his plans to provide a spectacular swansong for missiles like the US Minuteman or Russian SS-18 in the *Journal of Pyrotechnics*. It would make more sense to use these Cold War relics for entertainment rather than break them up for scrap, he argued.

As the missiles returned to Earth, they would release their cargo of thousands of artificial meteors, each weighing between 10 and 100 grams. The meteors would burn up in the atmosphere, with different colours depending on the chemicals with which they had been doped – sodium for yellow, strontium for red, and so on. Those in the first wave would create sonic booms. The noise would alert spectators to the display. 'It all happens pretty damn fast,' said Caulkins. However, he admitted that he hadn't worked out a solution to what was likely to be the most serious objection to his proposal: the fear that his pyrotechnic millennium party could be used as a cover for a real pre-emptive nuclear strike.

Some people even tried to ensure God really was on their side ...

And the voice said ...

An apocryphal tale perhaps, but one of our reporters might have heard the Voice of God. Apparently, researchers working with high-power laser weapons in 1999 created balls of plasma above the New Mexico desert which, when pushed around by the laser beams, created pressure waves that sounded like voices. The US military allegedly named the project the Voice of God and set about developing its use as a psychological weapon of war. It seems that they hoped to place the plasma balls in the sky high above the enemy and then speak to them – telling them, among other things, that their war was unjust or they should return home to slaughter their first-born. The military denied the experiments (as, of course, it would) and scientists were sceptical about whether such a phenomenon was possible. So apocryphal it remains ...

⚛ Cold War technology on a plate

Hot, crunchy pizza delivered to your door? More often than not, what turns up is a tepid slab of dough, topped with cold congealed cheese. But for those pizza-lovers who are too lazy to budge from the sofa, there was a brief glimmer of hope in 1994, thanks to high-tech military engineering. Claude Hayes, a defence contractor in San Diego, converted a device built for the "Star Wars" Cold War defence programme into a tray that kept pizzas piping hot for more than an hour.

The technology was originally used in the US government's Star Wars project as a lining for satellites that would defend them from laser attack. The material was a mesh of graphite fibres impregnated with a substance which, when hit by a laser, reacted chemically, absorbing massive amounts of energy and protecting the craft from damage.

'Things were going well,' Hayes recalled. 'But then the Russians got friendly.' As funding fizzled out, Hayes realised that the same principles that kept military hardware cool could also keep pizza hot. The material he invented was used as a tray which absorbed heat as the pizza cooked on it. It would then slowly release the heat when removed from the oven and could keep pizzas hot for more than an hour.

The US Star Wars project, which was intended to destroy enemy missiles from space while still in flight, seems to have spawned more than its fair share of offspring.

⚛ Icy asteroids could power reactors in space

Wacky ideas are commonplace among space scientists, but in 1993 officials at the US Department of Energy came up with a scheme to top them all. Searching for new ways to justify the department's moribund programme to develop a nuclear reactor for use in space, they proposed capturing nearby asteroids and comets and towing them into Earth orbit.

According to an internal memo from the energy department, nuclear-powered processing plants in space would extract water from the asteroids. The water could be used to sustain humans in space and as a propellant for nuclear rockets, it said.

'Electric energy beamed to Earth from space would provide nearly pure electricity without global warming, acid rain, strip mining or the hazards of fission products anywhere near Earth, three dozen years in the future,' the memo argued. There are believed to be thousands of asteroids relatively close to Earth, with water ice making up a large chunk of each. Some asteroids may be dormant comets, which would contain an even greater proportion of water.

But had the idea gained ground, those running the programme might have had a tough time convincing the public that the project would not endanger humans on Earth. 'There would be the twin bugaboos of nuclear power and comets,' said John Pike of the Federation of American Scientists. 'You would have to be able to convince people that the comet wouldn't come crashing into Kansas. Since the beginning of history, comets have been portents of disaster.'

Presumably, by the time we are able to pull asteroids into Earth orbit, we'll have computers that can do all the difficult calculations for us while reassuring us that nothing can go wrong.

However, at the moment, we are still working on numerous incarnations of artificial intelligence ...

❊ Forgive me computer, for I have sinned

Feeling a stab of after-hours guilt? If so, Greg Garvey's Automatic Confession Machine designed in 1993 might have been perfect for you. Just imagine stepping up to the sleek, black booth with its neon cross and winking Christ, kneeling on the red velvet stool, pressing the 'amen' button, and electronically divesting yourself of your sins.

Garvey, an artist at Concordia University in Montreal, who was brought up as a Catholic, programmed a computer to guide users through a formal confession, strictly according to church doctrine, he claimed. After the customary 'Bless me Father, for I have sinned', the programme instructed the user to tap in the number of days since last confession, the number of venial versus mortal sins (selected from a list of the seven deadly sins and the ten commandments) and then further details of each sin.

The confessor responded to the computer's questions by pressing buttons such as 'days' 'weeks', 'yes', 'no', 'venial' and 'mortal'. Once the sinful tally was recorded, the machine calculated penance and delivered a print-out for the penitent on a handy, wallet-sized slip of paper.

'My purpose is not to make fun of a particular sacrament or religion,' said Garvey. But while he admitted it was a somewhat tongue-in-cheek undertaking, he was quick to point out its possible advantages over the real thing. 'If we mass-produce these we can guarantee uniform application of the doctrine, we can automatically download the latest papal bull – and add any update necessary,' he said. 'We can have paid-for-confession – take MasterCard, Visa, American Express. And, of course, there's the added confidentiality.'

Garvey pointed out that his confessional was not the first marriage of high technology to religious rite. American televangelist Robert Tilton claimed to heal watchers who placed their hand on the television screen over his hand. And Jews had been faxing God for some time, in the form of prayers that were received in Jerusalem and then slipped into the cracks of the Wailing Wall.

But, said Garvey, the automatic confessional could not possibly replace a priest, because the software had not yet been ordained.

Trial by laptop

Imagine the scene: there's been a minor car crunch on a city street in Brazil, and the two drivers are arguing angrily over who's to blame and who should pay for the damage. Suddenly, a van screeches to a halt and out pop a judge, a court clerk and a very special laptop computer. Instant justice has arrived, cyber-style.

This is no fantasy. In the year 2000, a laptop was invented which ran an artificial intelligence program called the Electronic Judge, and its job was to help the human judge on the team swiftly and methodically dispense justice according to witness reports and forensic evidence at the scene of an incident. It could issue on-the-spot fines, order damages to be paid and even recommend jail sentences.

The software was tested by three judges in the state of Espirito Santo. It formed part of a scheme called Justice-on-Wheels, which was designed to speed up Brazil's overloaded legal system by dealing immediately with straightforward cases. The idea was not to replace judges but to make them more efficient, said Pedro Valls Feu Rosa, a judge in the state's Supreme Court of Appeals.

The E-Judge program presented the judge with multiple-choice questions, such as 'Did the driver stop at the red light?' or 'Had the driver been drinking alcohol above the acceptable limit of the law?' These are the sorts of questions that human judges are normally expected to answer, based on evidence from the scene, explained Feu Rosa, and they only need yes or no answers. 'If we are concerned with nothing more than pure logic, then why not give the task to a computer?

Most people were happy to have the matter sorted out on the spot, he said. The program gave more than a mere judgment: it also printed out its reasoning. If the human judge disagreed with the decision it could simply be overruled, said Feu Rosa. He admitted, however, that some people who had been judged by the program didn't realise that they'd been tried by software.

But, of course, whatever mind-boggling inventions and whatever great ideas scientists and researchers may come up with, they always have to be aware of the health and safety implications. Sometimes, it seems, scientists – as well as the public – need to be protected from themselves.

❊ Read this first

In 1996, reader John Isles wrote to tell us about information supplied with his Kenner Toy Company's 'Batman Returns' costume. 'Caution,' it announced. 'Cape does not enable user to fly.'

2 Mad research

There's nowt so queer as folk, runs the old Yorkshire saying. Well, the author is a Yorkshireman who used to agree with this axiom. And then he discovered scientists. If Yorkshiremen are a breed apart, in the way that Queenslanders, Texans and Bavarians all boast their superiority (to the irritation of their compatriots), they'll all be sad to learn that they are easily displaced by scientists at the top of the smugness ledger. Indeed, you'll see from this chapter that scientists are the black sheep among the infrequently spotted Leicester Longwools living on the organic rare breed farm's top pasture – very haughty, very headstrong (and sometimes wet and bedraggled). However much barbed wire you put around the fence, however deep the cattle grid, however secure the padlock on the only gate, come springtime you'll find the whole flock has chewed its way through the hedgerow, trotted down to the bottom field and got giddy by drinking undiluted sheep dip.

Working on the principle that you have to try everything once, the researchers here are all prepared to put their bodies on the line (more than metaphorically in some cases) in order to prove something. Or to fail to prove something, which is much the same thing. This means you'll encounter Jorge Mira Pérez and Jose Viña, who've gone to the trouble of calculating the temperatures of both heaven and hell; Konstantine Raudive, who probably failed to contact the dead, although he might deny it; and Troy Hurtubise, who really did put his body on the line when he donned a bear suit and went off to cavort with grizzlies.

Sadly, there are a few areas of 'research' reported by *New Scientist* over the years that we were unable to include. In an article on the effects that alcohol can have on human behaviour we recounted the story of the Polish farmer Krystof Azninski, who proved himself to be especially dim. Azninski had been drinking with friends, after which it was suggested they 'strip naked and play some "men's games"'. Initially they hit each other over the head with frozen swedes, but then one man seized a chain saw and cut off the end of his foot. Not to be outdone, Azninski grabbed the saw and crying 'Watch this then,' swung at his own head and chopped it off. 'It's funny,' said one companion, 'because when he was young he put on his sister's underwear. But he died like a man.'

We've since been told the story may be of doubtful authenticity. If it is, it was too good to avoid telling again. If it isn't, we apologise to Mr Azninski's family for reawakening their distress, but we are sure they'll agree that he deserves a mention in any scientific discussion of irrationality and is worth his place in the introduction to this chapter, if – sadly – not the chapter itself. Had Mr Azninski been a scientist conducting research it would have earned him a spot in these annals. But it seems that not even scientists would go to such lengths to prove alcohol gets you really, really drunk. So the Polish farmers merely warrant a footnote in our exploration of the outer reaches of scientific endeavour.

There's only one place to start when it comes to quality research. We're starting at the top, with God and the afterlife ...

☢ Too damned hot

In 1998, Biblical scholars breathed a sigh of relief when two physicists and a bishop decided that the furnaces of hell are indeed hotter than heaven.

In 1972, science waded into the domain of theology with an anonymous article in *Applied Optics* stating that heaven must be hotter than hell. The paper noted that Revelation 21:8 describes a lake in hell 'which burneth with fire and brimstone'. For there to be such a lake, hell's temperature must be below the boiling point of sulphur, just under 718 kelvin at atmospheric pressure.

Meanwhile, Isaiah 30:26 describes the lighting in heaven, where 'the light of the Moon shall be as the light of the Sun, and the light of the Sun shall be sevenfold, as the light of seven days'. Using Stefan's law, which states that the temperature of an object in thermal equilibrium is related to the fourth root of the amount of light it receives, the paper's authors calculated heaven's temperature to be a sweltering 798 kelvin.

However, in a letter published in the magazine *Physics Today*, Jorge Mira Pérez and Jose Viña, physicists at the University of Santiago de Compostela in Spain, said that the *Applied Optics* authors misinterpreted the Isaiah passage, wrongly multiplying 7 by 7 to make the illumination in heaven 49 times as bright as that experienced by us on Earth.

After Bishop Eugenio Ramiro Pose of Madrid confirmed that only a single factor of 7 was intended, Pérez and Viña recalculated heaven's temperature as 504.5 kelvin – blisteringly hot, but probably cooler than hell. 'A lot of colleagues, joking with me, have said that they prefer to stay on Earth,' said Pérez.

In the realm of intangibles

Poking through scientific archives can be an enlightening process. In the early 20th century, Duncan MacDougall, an American physician, positioned a patient dying from tuberculosis, bed and all, on an enormous beam balance, and waited with scientific curiosity for the end. After several hours, the patient died and 'the beam dropped with an audible stroke hitting the lower limiting bar and remaining there with no rebound. The loss was ascertained to be three-fourths of an ounce.' Therefore, it seems, a departing soul weighs as much as a slice of bread.

Of course, there's great debate as to whether a deity exists or otherwise, but as scientists we are on firm ground when we tackle more tangible issues like little green men. As this piece which appeared in New Scientist *in 1991 shows, we even know what they'll look like. And it won't be Vulcan.*

How to design an alien

What are the rules of thumb that life forms anywhere in the galaxy would have to follow? The famous geneticist Conrad Waddington believed that any higher life form would have to look like … Conrad Waddington. But most people see evolution as a contingent process – in other words, if evolution on Earth was run through again, land vertebrates – and that includes us – would be unlikely to reappear. And if they did we'd look very different. Of course, this applies to other planets too.

So, if we can't have humans on other planets, what can we have? There are patterns of general problems, and common solutions, that apply to life anywhere in the universe. We know this because different species on Earth invent identical solutions separately. Birds, bats, insects and some fish all fly. And plants and some bacteria photosynthesise. These universal solutions will be found on pretty well all other planets with life – including intelligence. So life will be formed of universal solutions – such as the elephant's huge legs to support great bulk in gravity – and local or parochial ones – such as its trunk, which developed from a need, on Earth, to pick up food from its feeding spots. Its food on another planet might not have required trunk-to-mouth delivery.

The difficulty, therefore, is in recognising universal solutions – which aliens will possess – and parochial ones – which they will not. Parochials normally happen only once – the trunk – universals more times – flying. Joints seem universal; the number of digits on a limb is not. Eyes, yes, external ears, probably not. And the list is very diverse … our strange excitement in sexual guilt pleasures is almost certainly parochial, so alien pornography seems unlikely.

This means the standard clichés of science fiction would not hold up. *Star Trek*'s Mr Spock, whose anthropomorphic appearance and evolutionary convergence is so close to humans that he can interbreed with us when we cannot even breed with species on our own planet, is, sadly, illogical. And we should disbelieve all the flying saucer stories that have little green men, not because they are little and green, but because they are men. Little green splots are so much more believable.

To be fair, little green splots are serious science. Communicating with the dead, however, requires taking a step towards the cliff edge of crumbling rationality …

❄ Night of the living-impaired

In 1998, the trade magazine *Computer Technology* ran an illuminating article by Michael Doherty about the technology of communicating with the dead.

The piece was full of fascinating information about what the author called 'necrophony'. Did you know, for example, that Thomas Edison attempted to construct a 'spirit communicator' in the 1920s? He didn't get very far but, undaunted, continued the work long after his death, communicating his progress through a medium called Sigrum Seuterman in 1967.

Since Edison's pioneering efforts, methods of communication with what we are obliged to call 'the living-impaired' have come on apace. Tape recorders, in particular, proved to be immensely useful in recording otherwise inaudible spirit voices. The late Konstantine Raudive, a leader in this field, compiled a collection of 72,000 tapes of spirit recordings while he was still with us. After his death, like Edison, he continued his work from 'the other side' and was himself recorded by his followers.

Modified telephones also played a part in necrophony. Several inventors came up with devices that enabled people to phone deceased relatives and friends. Notable among these was the 'spiricom', which Doherty described as 'a complex 29-megahertz communications system that established "quality" two-way conversations for the first time'. Unfortunately, a slight hitch with the spiricom emerged after the death of one of its inventors, William O'Neil, who related through the device that experiments on his side were infuriatingly being conducted at 68 megahertz.

The next step was obvious, and back in 1998 people were already working on it. It was hoped that we would be able to download software that would enable us to communicate with the dead through our computers. If Doherty was to be believed, at the time of writing several companies in the US were on the brink of releasing such software.

We foresaw problems when this article was published, though. How many dead people were computer literate? And would computers on the other side have been IBM-compatible, or would the living impaired prefer Apple Macs?

Rational or not, one of New Scientist's *correspondents had his interest piqued ...*

⚛ Rolling out Hades

We are disappointed that you see fit to make light of our latest research on communication between the living and the living impaired. At this very moment we are rolling out Hades Explorer 4.0, heralding an unparalleled increase in the volume of living-dead communications. You will be communicating with us soon, one way or the other.

> *Thomas Edison*
> *Necrophony Inc.*
> *Hades*
> *(17 January 1998)*

Is there really an afterlife? And if you are an egghead in this reality will you continue to be one after you've passed into that great lab in the sky? We hope so. But first let's deal with corporeal eggheads. There are, after all, plenty of them. In 2007 Alex Boese, author of Elephants on Acid and Other Bizarre Experiments, *went in search of them for* New Scientist *and stumbled on some of the craziest research ever. As he pointed out at the time, these researchers weren't cranks, all the work was carried out by honest, hard-working scientists who were not prepared to accept*

common-sense explanations of how the world works. All the
same, bet you can't help thinking 'eggheads' as you read about
them.

✳ Eyes wide open

Some people can sleep through anything. Earthquakes, gun-
shots, bright lights – nothing rouses them. But these are
people who are already asleep. In 1960, Ian Oswald of the
University of Edinburgh, UK, wondered how much stimulus
someone could be exposed to while awake and still drop off.
Would it even be possible to fall asleep with your eyes open?

Oswald first asked his volunteers to lie down on a couch.
Then he taped their eyes open. Directly in front of them,
about 50 centimetres away, he placed a bank of flashing lights.
No matter how much they rolled their eyes, they could not
avoid looking at the lights. Electrodes attached to their legs
delivered a series of painful shocks. As a finishing touch,
Oswald blasted 'very loud' music into their ears.

Three young men volunteered to be Oswald's guinea
pigs. In his write-up, Oswald praised them for their fortitude.
Yet, all the lights and noise and pain made no difference.
Once they were tired, an electroencephalograph showed all
three men to be asleep within 12 minutes. Oswald worded his
findings cautiously: 'There was a considerable fall of cerebral
vigilance, and a large decline in the presumptive ascending
facilitation from the brain-stem reticular formation to the cerebral cortex.' The men themselves were more straightforward. They said it felt like they had dozed off.

Oswald speculated that the key lay in the monotonous
nature of the stimuli. Faced with such monotony, he suggested, the brain goes into a kind of trance. That may explain
why it's easy to doze off, even in the middle of the day, while
you are driving along an empty road. How much this will

help when sleep eludes you while you're stuck on a red-eye flight is another question. Asking the baby in the row behind you to scream more rhythmically is unlikely to do the trick.

✳ Slumber learning

In the summer of 1942, Lawrence LeShan of The College of William & Mary in Williamsburg, Virginia, US, stood in the darkness of a cabin in an upstate New York camp where a row of young boys lay sleeping. He intoned a single phrase, over and over: 'My fingernails taste terribly bitter. My fingernails taste terribly bitter ...' Anyone happening upon the scene might have thought LeShan had gone mad, but he had not. The professor was conducting a sleep-learning experiment.

All the boys had been diagnosed as chronic nail-biters, and LeShan wanted to find out if nocturnal exposure to a negative suggestion could cure them. Initially he used a phonograph to faithfully repeat the phrase 300 times a night as the boys lay sleeping. One month into the experiment, a nurse discreetly checked their nails during a routine medical examination. One boy seemed to have kicked the habit. LeShan remarked that skin of a healthy texture had replaced the 'coarse wrinkled skin of the habitual biter'.

Then, five weeks into the investigation, disaster struck. The phonograph broke. Faced with having to abandon the experiment, LeShan began standing in the darkness and delivering the suggestion himself. Surprisingly, direct delivery had greater effect. Within two weeks, seven more boys had healthy nails. LeShan speculated that this was because his voice was clearer than the phonograph. Another possibility would be that his midnight confessions thoroughly spooked the children. 'If I stop biting my nails,' they probably thought, 'the strange man will go away.'

By the end of the summer, LeShan found that 40 per cent of the boys had kicked the habit, and concluded that the sleep-learning effect seemed to be real. Other researchers later disputed this. In a 1956 experiment at Santa Monica College in California, William Emmons and Charles Simon used an electroencephalograph to measure the brain activity of subjects, making sure they were fully asleep before playing a message. Under these conditions, the sleep-learning effect disappeared.

☀ The masked tickler

In 1933, Clarence Leuba, a professor of psychology at Antioch College in Yellow Springs, Ohio made his home the setting for an ambitious experiment. He planned to find out whether laughter is a learned response to being tickled or an innate one.

To achieve this goal, he determined never to allow his newborn son to associate laughter with tickling. This meant that no one – in particular his wife – was allowed to laugh in the presence of the child while tickling or being tickled. Leuba planned to observe whether his son eventually laughed when tickled, or grew up dismissing wiggling fingers in his armpits with a stony silence.

Somehow Leuba got his wife to promise to cooperate, and the Leuba household became a tickle-free zone, except during experimental sessions in which Leuba subjected R. L. Male, as he referred to his son in his research notes, to laughter-free tickling.

During these sessions, Leuba followed a strict procedure. First he donned a 30-centimetre by 40-centimetre cardboard mask, while as a further precaution maintaining a 'smileless, sober expression' behind it. Then he tickled his son in a predetermined pattern – first light, then vigorous – in order of armpits, ribs, chin, neck, knees, then feet.

   Everything went well until 23 April 1933, when Leuba recorded that his wife had made a confession. On one occasion, after her son's bath, she had 'bounced him up and down while laughing and saying, "Bouncy, Bouncy".' It is not clear if this was enough to ruin the experiment. What is clear is that by month seven, R. L. Male was happily screaming with laughter when tickled.

   Undeterred, Leuba repeated the experiment after his daughter, E. L. Female, was born in February 1936. He obtained the same result. By the age of seven months, his daughter was laughing when tickled.

   Leuba concluded that laughter must be an innate response to being tickled. However, one senses a hesitation in his conclusion, as if he felt that it all might have been different if only his wife had followed his rules more carefully. Leuba's tickle study does at least offer an object lesson to other researchers. In any experiment it is all but impossible to control all the variables, especially when one of the variables is your spouse.

❋ The look of eugh

Do emotions evoke characteristic facial expressions? Is there one expression everyone uses to convey shock, another for disgust, and so on? In 1924, Carney Landis, a graduate student in psychology at the University of Minnesota, designed an experiment to find out.

   Landis brought subjects into his lab and drew lines on their faces with a burnt cork so that he could more easily see the movement of their muscles. He then exposed them to a variety of stimuli designed to provoke a strong emotional reaction. For instance, he made them smell ammonia, listen to jazz, look at pornographic pictures and put their hand into a bucket of frogs. As they reacted to each stimulus, he snapped pictures of their faces.

The climax of the experiment arrived when Landis carried in a live white rat on a tray and asked them to decapitate it. Most people initially resisted his request. They questioned whether he was serious. Landis assured them he was. The subjects would then hesitantly pick the knife up and put it back down. Many of the men swore. Some of the women started to cry. Nevertheless, Landis urged them on. In the pictures Landis took, we see them hovering over the rat with their painted faces, knife in hand. They look like members of some strange cult preparing to offer a sacrifice to the Great God of the Experiment.

Two-thirds of the subjects eventually did as they'd been told. Landis noted that most of them performed the task clumsily: 'The effort and attempt to hurry usually resulted in a rather awkward and prolonged job of decapitation.' Even when the subject refused, the rat did not get a reprieve. Landis simply picked up the knife and decapitated the rodent himself.

With hindsight, Landis's experiment presented a stunning display of the willingness of people to obey orders, no matter how unpalatable. It anticipated the results of Stanley Milgram's more famous obedience experiment at Yale University by almost 40 years. Landis, however, never realised that the compliance of his subjects was more interesting than their facial expressions. He remained single-mindedly focused on his research topic. And no, he was never able to find a single, characteristic facial expression that people adopt while decapitating a rat.

✳ Terror in the skies

One day in the early 1960s, ten soldiers boarded a plane at Fort Hunter Liggett military base in California on what they thought was a routine training mission. The plane climbed into the clear blue sky, levelled out at around 5000 feet and cruised for a few minutes before suddenly lurching to one side as a propeller failed.

The pilot struggled with the controls and yelled frantically into his headset. Finally, he made an announcement over the intercom: 'We have an emergency. An engine has stalled and the landing gear is not functioning. I'm going to attempt to ditch in the ocean. Please prepare yourself.'

In such a situation, it would have been natural for the soldiers to feel fear or even terror. But there was no need. Though they didn't know it, they were in no danger. They were unwitting subjects in a study designed by the United States Army Leadership Human Research Unit near Monterey, California. Its purpose was to examine behavioural degradation under psychological stress – specifically, the stress of imminent death.

Having created a fear-arousing situation, the researchers next introduced a task to measure the soldiers' performance under pressure. The task was something most people find difficult under normal circumstances: filling out insurance forms. A steward distributed the paperwork, explaining it as a bureaucratic necessity. If they were all going to die, the army wanted to make sure it was covered for the loss.

Obediently, the soldiers leaned forward in their seats, pencils in hand, and set to work. They found the forms unexpectedly difficult to decipher, and quite likely they attributed this to the distraction of approaching death. In fact, the forms had deliberately been written in a confusing manner. They were, as the researchers put it, 'an example of deliberately bad human engineering'. Eventually the last soldier completed his form, and they all steeled themselves for the crash. At that point the pilot turned the plane around: 'Just kidding about that emergency, folks', and landed safely at the airfield.

Not surprisingly, anticipating a crash landing did interfere with the ability to accurately complete an insurance form. The soldiers in the plane made a significantly larger number of mistakes than did a control group on the ground who filled out the same paperwork.

Quite what the soldiers thought about their ordeal we don't know, but one of them did find a way to get even. When the plane next took off carrying a new group of subjects to terrify, the researchers discovered their experiment had been ruined. One of the earlier group had blown their cover by writing a warning message on his airsick bag.

What wags some of these eggheads can be. Aussie engineers, meanwhile, are a little more pragmatic, opting out of placing themselves in any danger and calling instead on feathered brethren. Their research is also an illustration of how much attitudes to animal suffering have changed in little more than a decade. Presumably artificial duck cadavers would be used in the caring 21st century.

⚛ Duck!

Ducks and other birds flying into the engines of low-flying aircraft are a hazard at many airports around the world. In 1995, *Gas Turbine News* reported that Australian engineers were researching the problem.

The difficulty they faced was how to simulate a duck flying into an engine to study what damage is caused and how it can be avoided. Their solution was not to simulate it at all. Instead they developed a gun that fired ducks at speeds of up to 273 kilometres an hour into their test engines.

They claimed that the ducks were humanely dispatched prior to being shot at the engines. We should hope so.

The ducks did not die in vain, as passengers on US Airways Flight 1549, which ditched with no loss of life in the Hudson River in January 2009 after a bird strike, will no doubt testify. Rather more prosaic, but still within the boundaries of egghead-ism, comes the kind of research that tells us how to live our mundane existence more easily. Vacuum cleaners, queuing at bus stops ... scientists can really improve your lot.

✳ Let it soothe as it sweeps as it cleans

In 1995, Frédérique Guyot, a PhD student at the acoustics laboratory of Pierre and Marie Curie University in Paris, made it her goal to identify what made the sound of a vacuum cleaner either intolerably irritating or pleasantly soothing. Armed with this knowledge, she hoped that manufacturers would be able to ensure that future models would make only the nicest of noises.

To find out what people hated most about the noise of their machines, Guyot recorded the sounds of 23 vacuum cleaners as they worked on a rug in an office. She played the recordings to 56 volunteers, who were instructed to imagine themselves pushing the vacuum cleaner around, and to describe their feelings about each sound. Machines were grouped into categories that ranged from 'best' through 'classical' to 'appalling'.

Guyot said that, unsurprisingly, loud vacuum cleaners were less popular than quiet ones. 'And ones that sound like a musical note – "hmmm" – are more unpleasant than those that just go "fffff",' she said. Least popular of all were loud appliances with high-pitched hums.

⚛ Lazy option is best

Ever lost patience waiting for a bus and decided to walk instead? Next time, stick around – it's nearly always the best strategy.

In 2008, Scott Kominers, a mathematician at Harvard University, and his colleagues derived a formula for the optimal time that you should wait for a tardy bus at each stop en route before giving up and walking on. 'Many mathematicians probably ponder this on their way to work, but never get round to working it out,' he said.

The team found that the solution was surprisingly simple. When both options seemed reasonably attractive, the formula advised you to choose the 'lazy' option: wait at the first stop, no matter how frustrating. The formula did break down in extreme cases, when the time interval between buses was longer than an hour, for example, and your destination only a kilometre away.

If you did choose to walk, you should make your decision before you start waiting, he said. You would still reach your destination later than the bus you'd have caught, but it would be much less frustrating than waiting for a while and then watching the bus shoot by. 'It certainly has changed the way I travel,' Kominers said.

Of course, missing a bus is annoying, and waiting for ever for one that doesn't come is probably even worse. But standing in line is a cinch compared to putting your body on the line. That's a different thing entirely.

✳ 'Bear-proof' suit to be put to the test

In December 2001, a Canadian man and a 3-metre, 585-kilogram Kodiak bear were set to face off, in an attempt to test a handmade, purportedly bear-proof suit.

The suit and its maker, Troy Hurtubise of North Bay, Ontario, won a 1998 Ig Nobel prize – awarded annually for improbable research projects – and an entry in the 2002 *Guinness Book of World Records* for the most expensive research suit ever constructed. Fifteen years of tinkering and US$100,000 went into the design, which incorporated plastic, rubber, chainmail, galvanised steel, titanium – and thousands of metres of duct tape.

The suit proved itself to be virtually indestructible. It survived two strikes with a 136-kilogram tree trunk, 18 collisions with a 3-tonne truck at 50 kilometres an hour, and numerous strikes by arrows, bullets, axes and baseball bats. 'I've never had a bruise,' said Hurtubise. But the suit had never come up against the very thing it was meant to protect against – a grizzly bear. On 9 December 2001, in an undisclosed location in western Canada, all that changed. In a 'controlled attack', the Kodiak, a larger, heavier subspecies of the grizzly, was to put it to the test.

The bear, which appeared on TV commercials and in movies, was to be instructed by its handler to attack for ten seconds. Showbiz aside, Hurtubise stressed that it was a real bear. 'Real teeth, real claws, real power,' he said. He fully expected the outside of the suit to be ripped to shreds. 'The suit's a toy to the bear,' he said. 'He'll make his way to the titanium.'

Hurtubise was banking on the titanium layers around the chest, head and lower body to protect him. 'If there's a weakness,' he said, 'it would be the chainmail joints.' Hurtubise said he was excited, but a little anxious too. 'Little things like trucks and baseball bats and axes and things – you don't feel that,' he said. 'This is a bear.'

So? What happened?

⚛ Bear-proof suit has surprise results

The first live tests of Troy Hurtubise's grizzly-proof suit in 2001 found that its best protective feature was its bizarre appearance. Hurtubise donned the suit and squared up to a 145-kilogram female grizzly but the bear just found it too weird. When confronted by Hurtubise in the Ursus Mark VI suit, the bear smelled a human, but saw an alien. 'There's no grizzly that's going to come near you in that suit,' the bear handler told him, after he spent ten minutes in a cage with the cowering animal.

The grizzly test was supposed to be the first live encounter. In an earlier test, the suit had to go it alone, without Hurtubise inside, with a fearsome opponent – a 545-kilogram male Kodiak. The suit was placed into the cage of the giant bear, to get him accustomed to it. Eventually, the bear began to sniff the contraption, and then proceeded to jump on it. 'You could hear the metal straining,' said Hurtubise. 'He started to investigate the thing with claw and tooth.'

Though it ripped off chunks of rubber, all was well until the bear began to shred the protective chainmail and was called off by the handler. Hurtubise learned that you should never skimp on chainmail. 'I should have used shark chainmail,' he lamented. 'Instead, I sent away for butcher's chainmail from France.'

Without effective chainmail, the bear handler decided not to allow an attack with Hurtubise inside the suit. But he did permit Hurtubise to take on the smaller female grizzly, although no contact was allowed. It took the bear a while to approach but soon enough it was just six inches away. 'I could feel her breath coming through my visor,' said Hurtubise. 'I was terrified.'

As far as we know Hurtubise never made it back into the cage with another grizzly. He reportedly sold the bear suit and invested the money in designing indestructible suits for soldiers and a device that allows humans to see through solid objects (as well as going bankrupt). Still, he's not the only researcher who ever donned an animal suit.

It takes one to know one

Some days when Joel Berger got dressed for work he put on a pair of jeans. On other days, it was a moose suit. That's right: a big, moulded styrofoam head covered in synthetic moose fur on top of a long moosey cape that reached down to his ankles. And if being the moose's front end sounds humiliating, imagine having to impersonate a moose's backside. That irksome task fell to Berger's long-suffering collaborator, Carol Cunningham.

'We do it to get data. It's that simple,' he said pre-emptively. 'We have to get close to unhabituated animals.'

In 1999, Berger and Cunningham, biologists at the University of Nevada in Reno, wanted to know if animals behaved differently when their predators disappeared. Moose in Wyoming, for instance, were at the top of the food chain. Over the previous decades, all their predators had died out, mostly through human interference. But it was a different story in Alaska, where wolves and grizzly bears still hunted the gentle beasts for supper.

So Berger and Cunningham wondered: how did the two populations differ? Did moose in Wyoming still react to the scents of their erstwhile predators? Or had they simply stopped bothering?

The question was not as easy to answer as you might imagine. Berger and Cunningham had to devise a way to put scents under a moose's nose and then measure how it reacted. But moose tend to live in remote forests and don't mingle casually with humans. First the researchers tried leaving dung and scent marks in the woods, going back later to see if the moose had approached them. That didn't work very well, said Berger. It could take weeks for the moose to run across the samples. 'By that time the odour is diffuse,' he said. He and Cunningham tried hurling the stuff in with masterful baseball pitches. They even catapulted it in with slingshots.

But in the end, going undercover as a moose seemed to be the only solution. Their unusual outfit didn't exactly keep them toasty – and the fur and foam had some major disadvantages when it came to outdoor pursuit. 'It's a hassle to use,' complained Berger. 'It's big and bulky. But it does work.' They even made moose sounds. 'It's a high-pitched "moo" – a "mew",' said Berger. He also pointed out that it wasn't exactly easy to navigate with the suit on, since they could barely see. The two had been known to fall over.

And a cumbersome moose suit was not their only distraction. Dangling round their necks were stopwatches, a pad of paper, pencils, a camera – and, er, research samples. 'Yeah, we're carrying bags of shit with us, too,' said Berger. Not just any old stuff, but Siberian tiger, grizzly bear, black bear and cat, not to mention an assortment of fine urines, including wolf, coyote and human. All in individual bags.

When they got to within about 25 metres of the moose, the real research would begin. Berger would make a 'scented' snowball and lob it near the moose subject. Sometimes she ignored it, said Berger. Other times, she ran away. A couple of times, moose got downright furious and looked as if they were about to charge. 'We dropped the moose suit and ran,' said Berger. 'We had to go back to get it.'

But if the moose did go over and sniff at the snowball, Berger and Cunningham would be ready to time how long it would take her to get back to foraging. They found that moose from regions where there were no predators took significantly less time to return to their feeding and were pretty uninterested, even when sniffing the faeces of ancient enemies.

We hope they keep it up. But we reckon that if Joel Berger had been a child he'd have been told that it was all neither funny, nor clever – but we think it is. However, we know what isn't: swearing. Rude words are neither clever nor funny. But studying the words and which people say them might just be ...

✳ Expletive deleted

Swearing 'recruits our expressive faculties to the fullest', wrote Harvard psychologist Steven Pinker in his book, *The Stuff of Thought*. Yet despite being a showcase for creativity, swear words are taboo in virtually all societies, even though their subject matter – usually sex or excretion – describes activities fundamental to human existence.

So why are we such potty-mouths, and what gives certain words the power to shock?

One theory is that cussing is the form of language that comes closest to a physical act of aggression. When you swear at someone, you are forcing an unpleasant thought on them and, lacking earlids, they are helpless to repel this assault. 'It's a substitute for physical violence,' said Timothy Jay, a psychologist at the Massachusetts College of Liberal Arts in North Adams. Most of us are able to restrain ourselves from launching these linguistic assaults – at least some of the time – but studies of people who lack this restraint are revealing.

Individuals with Tourette's syndrome have characteristic tics such as blinks and throat-clearing, and between 10 and 20 per cent also exhibit involuntary swearing, otherwise known as coprolalia. Diana Van Lancker Sidtis, a neurolinguist and speech pathologist at New York University, said that coprolalia can be regarded as a kind of vocal limbic tic. People with Tourette's have damage to a part of the brain called the basal ganglia – clusters of neurons buried deep in the front half of the brain that are known to inhibit inappropriate behaviour. This leads to uncontrollable swearing, she said.

As Pinker sees it, the basal ganglia are responsible for tagging certain thoughts as taboo. When the 'don't-go-there' label is no longer applied, as with Tourette's, taboo thoughts can reassert themselves and the urge to cuss becomes over-whelming. There is even one recorded case of a man with Tourette's who was deaf from birth and expressed his copro-lalia through sign language. In 2000, doctors at London's National Hospital for Neurology and Neurosurgery reported that his swearing was randomly interspersed within his signed speech, just as it is in other people with coprolalia. What's more, rather than flipping the finger or making other obscene gestures that hearing people deploy, he used the rec-ognised signs for rude words.

Tourette's aside, many people fear that bad language is on the increase. Most studies of the subject have found that men swear more than women. But a survey conducted in 2008 by Mike Thelwall at the University of Wolverhampton in the UK suggested that, among young British users of the social net-working website MySpace, strong swear words were used by males and females about equally – and a lot. 'Where once the only swear words young people wrote might have been fur-tively scribbled on the walls of public toilets, now they type them casually on to a computer screen,' said Thelwall. 'And there, they never run out of space.'

By poring over our rich library of filth, researchers have been able to get a handle on just what makes a good swear word. It is not just its sound, said Tony McEnery, a linguist at the University of Lancaster in the UK: after all, 'shot', 'ship' and 'spit' are not considered obscene, whereas 'shit' is. Besides, the German and French equivalents sound quite different and still pack a very satisfactory punch. It cannot just be about semantic content either, because the use of words denoting faeces or sexual matters in a medical context remains acceptable. 'Something about the pairing of certain meanings and sounds has a potent effect on people's emotions,' Pinker said.

Swear words also go in and out of fashion in line with the taboos they breach. '"Damn" was the undisputed king before "fuck" arrived on the scene,' said McEnery, 'but lost its piquancy as the fear of burning in hell faded.' 'Poxy', 'leprous', 'canker' and other disease-related words went the same way as hygiene improved. From the 19th century on, English speakers have mostly vented their frustration by reference to two different classes of taboo: the sexual and the scatological.

Today 'fuck' reigns supreme, but there is still room for innovation. So, what will be the next big thing in swearing? Most experts decline to predict any winners. 'Paedophile' seems to be an especially offensive thing to call someone today, explained McEnery, and therefore a good candidate, but there is no evidence it is gaining ground as a swear word. 'Something is missing with that word,' he said. Pinker said the word 'cancer', while not yet obscene, is acquiring taboo characteristics – note how it is often referred to as 'The Big C'.

Jay road-tested a few contenders of his own. He once muttered 'Expletive!' on a golf course, and got some strange looks. You can't impose swear words on a language, he noted, they arise organically. Just because you personally dislike cheese, shouting 'stilton' out loudly is unlikely to catch on.

Bloody hell, fancy that!

3 The yuck factor

Welcome to the world of yuck, or what the academic world calls grossology. Anybody who had anything to do with the creation of this book insisted that we include a chapter on the yuck factor. Which, as luck would have it, we had every intention of doing from the very beginning. How could we deny anybody the pleasure of reading about the smell quotient of farts?

We're not going to suggest that we came up with the science behind this chapter. But what we are delighted to discover is that grossology is a genuine scientific field practised by its leading proponent and inventor, Sylvia Branzei, who gave birth to the discipline in 1993 in order to get children more interested in biology and chemistry. But what exactly qualifies a subject to fall into the grossology category? To be honest, it seems to be everything that kids find funny. Farts are a very important component of the discipline. As is urine. And faeces. Which reminds us of an old joke.

Five-year-old Alice walked into the kitchen one morning as her biologist father was reading the newspaper. 'Where does poo come from?' she asked. Startled, he regarded her for a moment before replying. 'Well,' he began. 'You know we've just had breakfast?'

'Yes,' she replied. He went on: 'The food goes into our tummies and our bodies take out all the good stuff and whatever's left comes out of our bottoms when we go to the toilet, and that's poo.' She was silent as she digested the enormity of what he'd divulged. 'And Tigger?'

Naturally, you don't get far in the field of grossology without knowing a thing or two about underpants, vital if you are thinking of going into space. Let's face it, what on Earth – pun intended – do you do with a pair of dirty Y-fronts when you are 3 parsecs from the nearest solar system? You design a cocktail of bacteria to eat them, according to Russia's Institute for Biological and Medical Problems.

Yes, soiled underwear is big in the world of yuck. One set of smalls that didn't make it into this chapter, however, were the damp leather pair found in the wreck of a Bronze-Age boat being excavated near Dover in England. They were of great importance archaeologically, being possibly the world's oldest surviving pair, but we struggled to justify their inclusion in this chapter because they weren't so much gross as a vital component of the vessel they were found in. The researchers who discovered them said they had been used to plug a leak in the boat's hull during a Channel crossing. We wonder which sweaty oarsman volunteered to hand his over when the boat started listing?

Since its invention, grossology has splintered into different branches, most of which, down the years, New Scientist has covered in depth. First up, bodily excretions. We're not sure our lead story comes as much of a surprise, but science exists to confirm the expected as much as it does to discover the unexpected.

✳ It's a gas

Australian men fart more than Australian women, and their farts smell more. These were the findings of one of the largest ever studies of flatus emissions (what the wider world knows as farting), carried out in 1993 by R. A. Stanton, a nutrition consultant from Sydney, and T. D. Bolin of the Prince of Wales Hospital in Randwick, New South Wales.

Each of the 60 men and 60 women who took part in the study was given a small pocket 'Flatometer' to record the number of daily emissions. The participants also gave a subjective estimate of the aroma of the emission, rating it on a 4-point scale from 'odourless' to 'severe' in order to arrive at an 'aroma quotient' for the day's farts. A dietary diary was kept to reveal the flatus potential of different foods.

At the end of the 5-month study period, men reported an average 12.7 farts per day with a mean aroma quotient of 0.86. Women reported a lower 7.1 farts per day, with a more discreet aroma quotient of 0.54.

The study revealed a high correlation of emissions with dietary fibre intake, consistent with popular wisdom about baked beans. However, an expected correlation between farts and beer remained unproven.

We worry that this survey may suffer from the kind of objectivity failure that befalls similar surveys studying sex lives. Men tend to boast and women tend to be more circumspect regarding their sexual experiences (and/or aroma quotient) out of respect for social 'norms'. Nonetheless, we are certain there is a commercial opportunity for flatometers, because there's obviously a thriving research community out there studying farts …

Raising a stink

Here's a piece of trivia with which to impress your drinking pals. The average adult in the Western world farts roughly 10 times a day, releasing enough gas to inflate a party balloon. More than 99 per cent of these emissions are made up of five odourless gases. What exactly causes their foul smell has long been a matter of debate. But in 2001, one man believed he had the answer.

Michael Levitt, a gastroenterologist at the Veterans Administration Medical Center in Minneapolis, had been studying flatus for over 30 years and solved numerous mysteries that might otherwise have troubled the flatulent reader. Well known in gastroenterological circles for his painstaking research and undying enthusiasm, Levitt had written more than 200 papers on this subject and is acknowledged as a world authority. Articles about him and his work have appeared in everything from *The Sydney Morning Herald* to *The Daily Telegraph*.

Many scientists would welcome this exposure, but for Levitt it was disastrous. Readers would write angry letters to his employers complaining that his research was a waste of money. It became so bad that he eventually refused to talk to the press for fear of jeopardising his career. Just what went wrong?

Part of the problem was the mirth-provoking nature of the topic itself. The temptation to turn out stories brimming with puns and fart jokes was often too great for journalists to resist, and the tone of the story inevitably colours Levitt's image in the mind of the readers. *The Sydney Morning Herald*, for example, began its story about Levitt with the euphemistic headline 'Tail wind'. *The Washington Post* started cheekily with 'Waiting to inhale' but dampened further enthusiasm with the sub-heading 'Mainstream and alternative practitioners agree that diet is central to controlling flatulence'. And online magazine Salon.com plumped for 'Dr Fart speaks'.

Levitt's work attracted the media because it is serious research on a snigger-inducing topic. For instance, he was the first to correctly identify the gases that make farts smell. Previously, researchers had suspected that the guilty parties were foul-smelling compounds such as indole and skatole, created by the breakdown of amino acids in the gut. But nobody had ever bothered to check.

Evaluating a smell is a difficult task, so Levitt turned to the noses of two people with a rather unusual ability. Both could identify different sulphur-containing gases purely by smell. These lucky individuals were asked to evaluate the flatus of 16 healthy men and women who, the previous evening, had eaten 200 grams of pinto beans to ensure ample gas production. Levitt said the results pointed to hydrogen sulphide as the culprit in smelly farts, accompanied to a lesser extent by other sulphur-based gases. But indole and skatole were nowhere to be smelt.

Levitt even tested an artificial detox system – a commercially available device claimed to be able to reduce the odour of farts. The Toot Trapper was a foam cushion covered on one side with activated charcoal – charcoal with an increased surface area – which absorbed certain gases. Since no standard technique existed to carry out such tests, Levitt developed his own by designing airtight Mylar pantaloons to trap the gases for analysis. The good news is that the Toot Trapper worked well, reducing the concentration of sulphur-based gases by a factor of 10. The bad news is that it was rather unwieldy.

Research like this was bound to attract media coverage. But Levitt's work was far from pointless – it has saved lives. Hydrogen and methane, two of the main gases that form in the gut, are combustible. In the 1980s, they caused a number of fatal explosions during otherwise routine operations on the gut. Somehow the purgatives used to clean the gut enhanced the production of hydrogen or methane and a chance spark during the operation triggered an explosion. Levitt and others have since developed purgatives based on polyethylene glycol, which clean the bowel with minimal gas production. Colonic detonations are now rare.

Levitt also studied a 28-year-old man who meticulously recorded every passage of gas he produced over a 3-year period. Farting about 34 times a day, he let out over 8 litres of gas, equal to the gas in several party balloons. Through Levitt, he achieved the dubious fame of being the only thoroughly studied man with excessive flatus on record.

Journalists certainly seemed to like Levitt, and stories about him were littered with the good-humoured quotes he would give. Explaining to *The Washington Post* that gut bacteria absorb gas as well as produce it, he said: 'If we passed all the gas that we made, everybody would be farting a million times a day.'

That's enough about farts. Time to move on to some research that's a bit more solid.

✳ Engage dark matter!

'I'm trying as hard as I can, Captain,' exclaims Scotty, the strain etching lines in his forehead. 'I can't give you any more!' But unlike *Star Trek*'s *Enterprise*, it was once hoped that spacecraft might use a less savoury energy supply than Scotty's beloved dilithium crystals: human waste.

In 1999, NASA enlisted the aid of Advanced Fuel Research of Connecticut in a $600,000 project to turn astronaut waste – either faeces or some plastics – into a power source for spaceships using pyrolysis: the breaking down of waste by heating it in the absence of oxygen.

Normally, when organic molecules such as those found in faeces or in plastic are burned, they combine with oxygen in the air, producing carbon dioxide and water. But in pyrolysis, there is no oxygen to combine with, so the molecules break their bonds and rearrange themselves into smaller molecules. According to AFR scientist Mike Serio, 'Things start breaking down at about 350 °C, and what you start making includes a lot of liquids. At 600 °C or 650 °C, you break down the liquids into gases. It does give you flexibility.'

You could burn these liquids or gases to release energy, or turn them into plastics or other organic materials. Pyrolysis could produce heavier molecules such as benzene or toluene, and could be a source of raw materials to make plastics or rubber. It could also create ammonia for fertiliser.

'You can use human waste as well as other waste, like scrap plastic bags,' said Jim Markham, AFR's chief executive officer. And you wouldn't have to worry about variations in the consistency and content of the waste material – the pyrolysis unit would be able to handle them all. 'It's tailored to unpredictable mixtures,' he said. 'Ideally, you'd dial in the desired outcome and it would compensate.'

Research into the usefulness of human waste products might be relatively new, but turds themselves have been with us for a long time. Long enough to become part of the fossil record. Apparently, the oldest terrestrial faeces discovered were found by researchers at the University of Wales in Cardiff. British academia leads the world once again. Believed to be from the earliest land-dwelling herbivores, the turds are 412 million years old. There's probably not too much you can learn from them once they've turned into rock, but the same cannot be said about whale faeces.

✳ Sniffing out whale poo

'I hope we get a poop today,' said Rosalind Rolland, a conservationist at the New England Aquarium, pouring her morning tea. 'Just one.'

'Maybe two,' said her colleague Scott Kraus. 'If you follow something long enough, it'll poop.'

Their colleague Fargo, having dedicated over half his life to the subject, was even more enthusiastic. In 2006, Fargo was one of an elite corps of whale-scat sniffer dogs, and perhaps the most important member of the research team.

Whales may be the biggest creatures on the planet, but they also swim fast and dive deep, so are notoriously difficult to study. Fortunately, you can tell a lot about an animal by studying what it leaves behind. In faeces, researchers can find clues about an animal's distribution, abundance, sex and reproductive status, stress levels, health and vitality.

Spotting scat samples, which hang just below the sea's surface before they break apart, is far more difficult than finding the animals themselves, and that's where dogs like Fargo come in. With years of training and more than 200 million olfactory receptors – compared with humans' mere 5 million – dogs can ramp up the number of samples.

Upwind of a few whales, Fargo dropped his snout. 'He's got scent' said Rolland. Fargo paced back and forth across the bow but then stopped moving. He had lost the scent. 'It might have been a big fart,' Rolland conceded.

Rolland figured that faecal sampling would help her narrow down why right whales had stopped reproducing in the North Atlantic in the late 1990s. She thought faecal samples could be used to test reproductive hormones, as had been done in primates and other animals. They began with just humans searching for whale poo, but when dogs were introduced samples increased fourfold. On average, dogs collected about a scat an hour – a rate Rolland referred to as 'poops per unit effort'.

Whale faeces are surprisingly modest, the size of a small brick and a reddish brown. When the team finally located one, research assistant Cindy Browning scooped it up with a fine-meshed plankton net. Rolland and Browning carefully stored the sample in palm-sized plastic jars. Once it was brought on board, the smell was overwhelming. 'If you spill it on your clothes,' Rolland warned, 'you want to throw those clothes away.'

In 2006, the samples were proving invaluable. Rolland and her colleagues showed that paralytic shellfish poisoning, caused by eating shellfish contaminated with toxic algae, may have contributed to the right whale's failure to recover from centuries of hunting. Perhaps even more troubling was the presence of domoic acid in right whales. This toxin has several neurological effects, including seizures and coma, and had been responsible for the death of hundreds of sea lions in the Pacific. Protozoan parasites could also have caused problems. These had been associated with diarrhoeal disease in land mammals, and it was thought that the blame might have lain with humans. Right whales spend much of their lives close to big cities. They might have been picking up these parasites from human or domestic animal waste dumped in the ocean.

Hormone studies, by contrast, brought some good news. By looking at levels of progestogens and oestrogens, researchers were able to estimate the reproductive status of females, and many were doing well. The study had also been expanded to study killer whales, whose faeces, for the record, are a snotty greenish brown. They are also less buoyant than right whale scat and difficult to spot from a boat. Once back in New England, Rolland reflected on the work still to be done to protect the whales. 'I thought I'd be into something else now,' she said, 'But I've just gotten deeper into it.'

With so much poo about, you'd be forgiven for thinking there would be no entrepreneurial benefit in manufacturing extra. You'd be wrong.

⚛ Faking it

In 1994, Richard Yeo and Debra Welchel of the Kimberley-Clark Corporation in Dallas, Texas, invented a recipe to simulate something you might have thought was fairly abundant in America: human faeces. Their mix consisted of polyvinyls, starches, natural gums and gelatins, all water-soluble, plus insoluble fibres and resins. Add water to this and you would get a synthetic turd 'as close to the real thing as possible', Yeo told the *New York Times*.

This is impressive, but by now a question may well have occurred to you: why would we need it? The answer was that Kimberley-Clark makes nappies and adult incontinence pads and these products must be rigorously tested. 'But the technicians have some objection to handling the real thing,' Yeo said, adding that actual faeces are 'a form of biologically hazardous material'.

In addition, the firm's labs might have found they needed, say, 10 kilograms of faeces at 9 o'clock on a Monday morning, to begin tests on a new range of nappies. The real thing, Yeo complained, can be 'a bit difficult to obtain, even from infants'.

Any scientific endeavour typically starts with a review of the existing literature. 'We had fundamental studies of real faeces,' Yeo said, 'so we knew what was needed.' Previous attempts to make do with peanut butter or pumpkin pie mix had failed – their liquids and solids separated too rapidly.

The material mimicked faecal behaviour far more accurately, in all but one respect: it had no odour.

Of course, fake poo is only half the story. Some people are making fake urine too. Just what is it about Texans and phoney human waste?

❋ Just add water

In 1986 in Austin, Texas, entrepreneur Jeff Nightbyrd marketed instant urine for the growing number of employees being asked by their bosses to undergo tests to see if they were using drugs.

Like instant coffee, the urine came in powder form, and was reconstituted by adding hot distilled water. It could be ordered by mail at a cost of $19.95 for 2-gram vials, providing enough urine for two samples. Nightbyrd never said how he turned urine into powder, but it must have been a very smelly business.

And from faking it, to reusing it. 'Waste not, want not' is a maxim held in high regard by interstellar travellers. In space, no one can hear you pee …

❋ Purified urine to be astronauts' drinking water

In 2008, as NASA prepared to double the number of astronauts living aboard the International Space Station, nothing was expected to do more for crew bonding than a machine being launched aboard the space shuttle *Endeavour*: a water-recycling device designed to process the crew's urine for communal consumption.

'We did blind taste tests of the water,' said NASA's Bob Bagdigian, the system's lead engineer. 'Nobody had any strong objections. Other than a faint taste of iodine, it is just as refreshing as any other kind of water. I've got some in my fridge,' he added. 'It tastes fine to me.'

Delivery of the $250 million wastewater recycling gear was among the primary goals of NASA's 124th shuttle mission, which launched from the Kennedy Space Center in Florida in November 2008. In addition to the water recycler, the shuttle *Endeavour* also carried a second toilet.

'With six people, you really do need to have a two-bathroom house. It's a lot more convenient and a lot more efficient,' said *Endeavour* astronaut Sandra Magnus. Reusing water will become essential once NASA retires its space shuttles, which produce water as a by-product of their electrical systems. Rather than dumping the water overboard, NASA has been transferring it to the space station.

But the shuttle's days are numbered. NASA is preparing to end the programme in 2010, after which Russian Soyuz spacecraft will be the only way to ferry crew to the space station. 'We can't be delivering water all the time for six crew,' said space station flight director Ron Spencer. 'Recycling is a must.'

NASA expected to process about 6 gallons (23 litres) of water per day with the new device. The goal was to recover about 92 per cent of the water from the crew's urine and moisture in the air. The wastewater was processed using an extensive series of purification techniques, including distillation – which is somewhat tricky in microgravity – filtration, oxidation, and ionisation. The final step was the addition of iodine to control microbial growth. The device was intended to process a full day's worth of wastewater in less than 24 hours. 'Today's drinking water was yesterday's waste,' Bagdigian said.

The first recycled astronaut urine came on stream in the International Space Station in May 2009. Apparently it tastes 'great'. But while urine can seemingly be used again and again, it appears that dirty underwear is not for sharing.

✳ Less waste, more space

One of space travel's most pressing but least-known problems is what to do with dirty underwear. One solution, attempted by Russian scientists in 1998, was to design a cocktail of bacteria to digest astronauts' cotton and paper underpants. They had hoped that the resulting methane gas could be used to power spacecraft.

The disposal unit would be able to process plastic, cellulose and other organic waste aboard a spacecraft. Cosmonauts identified waste as one of the most acute problems they face. Each astronaut produces an average of 2.5 kilograms of uncompressed waste a day. To keep waste to a minimum, they are forced to wear underwear for up to a week, and discarded undergarments are burned up in the Earth's atmosphere in waste modules that call twice a year.

The researchers aimed to complete their microbial disposal unit by 2017 to coincide with Russia's planned launch of a crewed interplanetary mission to Mars.

It's either that, or the cosmonauts pay a visit to Alex Fowler. If they did, they might be able to wear their smalls indefinitely ...

✳ No more laundry

Most of us try to wash bacteria out of our clothes. Not so Alex Fowler, who wanted several thousand bugs to set up home inside every single fibre of a fabric, living, breeding and eating up the dirt. Welcome to the world of self-cleaning clothes.

Eventually, the garments in your wardrobe may be able to support a variety of bacteria engineered to eat odour-causing chemicals and human sweat. Other bacteria might secrete waterproof and protective coatings to extend the life of clothing and produce antiseptic for bandages. Ironically, textile makers have spent millions developing fibres blended with, say, silver ions or chlorine to kill off the bugs in fabric.

But encouraging bacteria to grow on fibres turned out to be harder than Fowler had expected. 'I thought they would wick into the fibre by capillary action, but it didn't turn out like that,' he said. Brute force was required. In 2001, Fowler and his team from the University of Massachusetts at Dartmouth developed a vacuum pump that could connect to the end of hollow fibres from the milkweed plant. Although no longer used to make clothes, the plant is still used to make rope.

The pump sucked a few drops of agar jelly containing *Escherichia coli* into the fibre. The bacteria easily formed a thriving colony and began to breed. After some initial tests, Fowler had no problem firing several hundred bacteria into the fibre. 'They're tough little guys,' he said. Fowler used a harmless strain of *E. coli* genetically engineered to produce a fluorescent protein from a jellyfish to make the bacterium glow as it grew, allowing researchers to monitor its progress.

The group weren't sure how long the bacteria could survive in the fibres. They suspected they might become dormant after several weeks when their food supply ran out, but Fowler hoped to reactivate them by soaking the milkweed fibres in additional nutrients. So if your shirt was impregnated with a strain of *E. coli* designed to feed on human sweat and the proteins that cause body odours, you'd only have to wear it to jolt the bugs into action. For some other strains, you might have to douse it with additional nutrients occasionally. 'You could end up having to feed your shirt instead of wash it,' said Fowler.

All this orbital recycling of urine and bacterial underwear destruction raises a question though. In zero gravity, just how do you emit all the waste?

❋ Wind in the loo

From the outside, the 1972 space shuttle toilet looked much like the uncomfortable apparatus found in the jet airliners of the era. Inside, however, liquid and solid wastes were directed into separate pipes by high-velocity air streams that compensated for the lack of gravity on board the shuttle. The waste was held in two tanks – solids being vacuum-dried, sterilised and deodorised, ready to be pumped out when the shuttle returned to Earth. Depositing them in orbit would have obscured rearward vision and possibly interfered with external equipment.

Very clever, but not as ingenious as Göran Emil Lagström's multipurpose device …

❋ Hot seat cleans up

Most of the best inventions kill at least two birds with one stone. Göran Emil Lagström's 1972 invention both disposed of lavatory waste where there was no proper sewage system and provided central heating. The contraption consisted of a combustion chamber with an oil or gas burner to burn solid waste and evaporate liquid waste, all situated under the toilet pan and surrounded by a water jacket. The jacket was connected to a large column-like tank which fed a central heating system.

Inside the tank, and heated by it, was a second, completely separate water-heating system which provided hot water for washing. The water was supplied from the mains or a water store.

The hot exhaust fumes from the burning waste were passed from the combustion chamber to a chimney.

Lovely. But enough about toilets. Time to move on to snot, bad breath and mouldy food. In 2002, New Scientist *ran this article about the mysteries behind the runny nose.*

⚛ Season's sniffles

SCHNIFFFRRPGHRRT … Ah, that's better. Uh oh, unbecoming drip is rapidly re-forming on end of nose. Hunt frantically through pockets for bedraggled scraps of paper hanky. No joy … oh well, only one thing for it. Time to deploy the sleeve.

Admit it, you've been there. You are fabulously dressed to impress, ready to mingle, frolic, see and be seen. After a 20-minute walk in the freezing cold, you arrive triumphant. Then, as your body warms up and you begin an intelligent conversation with your boss, your nose begins to drip like a leaky faucet. The phenomenon has earned an entry in medical textbooks as 'cold-induced rhinitis'. But don't be fooled by the fancy name. The truth is that no one knows very much about it, although scientists have pointed the finger at the autonomic nervous system, which directs involuntary actions such as breathing. Nerves belonging to the autonomic nervous system, some of which connect to the nasal glands, use a neurotransmitter known as acetylcholine. Fortunately, there is a quick fix. Ipratropium bromide is an inhibitor of acetylcholine. Two squirts of the solution in each nostril 45 minutes before skiing, snowballing or eating spicy food and the problem is taken care of.

Most people go through life unaware how easy it is to fix the problem. According to William Silvers, an allergist from Vail, Colorado, whose fascination with runny noses won him recognition as 'The Skier's Nose Doctor', people don't do much about their sudden bouts of sniffling because they are simply not seen as a medical issue. In fact, Silvers claimed, when faced with the grim realisation that their nose has begun to leak, the majority of folks take the matter into their own hands – quite literally. 'That's why they make cycling gloves with absorbent terry cloth on the outside,' he said. You thought they were designed to mop sweat? 'Well, I use them for my nose,' he replied, unabashed.

Silvers has been researching cold-induced runny noses since the 1980s, using the slopes of several Colorado ski resorts as his field lab. And his findings on people's nose habits make intriguing reading. Women tend to be better prepared and often carry tissues, for example. Men prefer their sleeves, or in certain situations (especially in the great outdoors) may even resort to what he refers to as 'the farmer's blow': block one nostril and blow, then repeat for the other side. 'They blow it out and keep on going,' said Silvers. Children just ignore it. Wiping may be their parents' favoured option, but a good old snort to recycle that lovely snot is good enough for the kids. 'It doesn't bother them as much,' he said.

Silver's fascination with the runny nose doesn't end there. He could even tell you how common the phenomenon is based on quantifying the usage of tissues by placing boxes of them near ski lifts. The colder it is on the slopes, he found, the faster the tissues disappear.

A few teams around the world have carried out studies of cold-induced rhinitis in animals as well as humans. But the most fiendish snot-inducing experiments have involved people. At Johns Hopkins University School of Medicine in Baltimore, for example, volunteers were asked to inhale cold dry air for 45 minutes so that scientists could measure their nasal emissions in search of any hint of why the phenomenon is different from, say, an allergic reaction. There have even been studies in which just one nostril was blasted with chilly air. Surprisingly, both nostrils were equally prolific producers of mucus in this situation.

Of course, plenty of people have their own pet theories as to what causes the problem. The ancient Greeks thought that a runny nose was a sign that the brain was turning into mush. One day, science may reveal just why copious amounts of watery goo end up on our sleeves when we least expect it. Until then, only the nose knows.

And if they can collect whale poo, why not whale snot too?

⚛ Thar she blows

What is the strangest thing you could do with a remote-controlled toy helicopter? Strapping on a few Petri dishes and flying it through whale snot must be high on the list. That is how Karina Acevedo-Whitehouse, a veterinarian and conservation biologist with the Zoological Society of London, has spent much of her time over the past few years.

In 2008, Acevedo-Whitehouse made it possible to study the viruses, fungi and bacteria that hitch a ride in whale lungs for the first time.

Researchers can fairly easily take blood samples from other marine mammals, such as seals and sea lions, but a whale's sheer bulk means that such a sampling would be fatal. 'Scientists have always found it difficult to study diseases in whales because of their size,' explained Acevedo-Whitehouse. But after witnessing the sheer power of whale 'blows' in the Gulf of California, she realised that this would be the best way of sampling the insides of a live whale in the ocean.

She first tried tying herself to a research boat and leaning overboard to catch a bit of whale snot in Petri dishes. 'It worked,' she said, 'but it wasn't very safe.' Her technique became more sophisticated. For species like grey and sperm whales that did not mind being close to a boat, the researchers attached their Petri dishes to a long pole and held them out over the blows.

With the shyer blue whale, they had to resort to miniature helicopters. The Petri dishes were attached beneath the metre-long choppers, which were remotely flown through whale snot. 'The whales definitely notice the helicopter,' said Acevedo-Whitehouse, 'they turn on their sides to look at it. But they don't seem bothered and we don't even touch them.'

Fortunately, the researchers are not being forced to study the odour of whale blows. But if they did, they might be able to determine what sex they were. Halitosis has hidden secrets.

Bad breath points to sex

People can correctly guess the sex of fellow humans simply by smelling their breath, according to a study carried out in 1982 at the University of Pennsylvania in the US.

In their experiments, 19 female and 14 male college students acted as odour donors; they were asked to refrain from cleaning their teeth or eating garlic or heavily spiced foods during the five days of the study. The odour judges consisted of five male and five female students.

Odour donors and judges were separated by a screen in a large, well-ventilated room. Each donor exhaled a given amount of breath into a glass tube passing through the screen. On the opposite side, the judges inhaled the breath of a donor and ranked it on a seven-point scale according to its intensity and pleasantness or otherwise. In addition, each judge was asked to guess the sex of the donor.

Both male and female judges correctly guessed the sex of the donors in 95 per cent of tests. Female judges were better at identifying males and male judges slightly better at identifying females. On average, male odours were more intense and less pleasant than female odours. Circumstantial evidence points to hormonal differences.

❀ Real slow food

The next time you mourn a forgotten morsel that's slipped past its use-by date, remember that things could always be worse. Take the case of Fidelia Bates of Tecumseh, Michigan. After baking a fruit cake for Thanksgiving in November 1877, the unfortunate Mrs Bates promptly expired. This presented a rather delicate question at the family farmhouse: who would be the first to eat a piece of the dead woman's cake?

As it turned out, nobody would. Mrs Bates's family has resisted temptation for over 130 years, and counting. 'It's hard, it's crystallised, it's fossilised,' said her 86-year-old great-grandson Morgan Ford. 'Nobody wanted to eat it after she passed away, and so now I have it.' Kept under a glass lid and stored high up in a cupboard for 75 years, the cake has stayed there ever since, save for the occasional appearance on TV or at Morgan's grandchildren's school show-and-tell.

The fruit cake has attracted a few daredevil gourmands over the years. 'My uncle was the first to try a tiny piece off it, about 25 years ago,' said Ford. 'And I did lift the lid off when it turned 100 and for a moment we could smell rum.'

Such wizened leftovers have long been a staple of local newspaper reports. In 1951, Mrs E. Burt Phillips of West Hanover, Massachusetts, returned a 56-year-old can of clams to the manufacturer ('still edible', the press duly reported). A year later a 70-year-old crock of butter ('still white and sweet') was retrieved from an abandoned well in Illinois. In 1968, Sylvia Rapson of Cowley, UK, found a loaf of bread baked in 1896, still edible, tightly wrapped in table linen in an attic trunk ('I'm keeping it for sentimental reasons,' she informed *The Times*), and when a house in Grimsby was razed to the ground in 1970, the ruins miraculously yielded up a 1928 packet of breakfast cereal – a find that was declared, inevitably, 'still edible'.

More unusual is an original owner of ancient grub who's actually willing to eat it. In 1969, one George Lambert turned up at the New Mexico state fair wearing his uniform from the 1898 Spanish-American war. Inside his mess kit he found a piece of hard tack and to the crowd's awe he bit a piece off and ate it. 'Tastes just like it did then,' the grizzled veteran announced. 'Wasn't any good then and it isn't now.'

Oscar Pike, a food scientist at Brigham Young University in Provo, Utah, would probably not have been surprised. 'Food quality is always declining,' he said. Just how quickly it declines, though, remained anecdotal until Pike and colleagues put out a call to local households: bring us your tired, your stale, your undusted masses of tins and sacks. In short, empty those basement pantries.

Having conducted taste and odour tests on everything from 30-year-old dried milk to oatmeal, Pike announced his results in 2006: the oatmeal wasn't all that good, but not all that bad either. It helped that the fats in quick-cooking oats do not readily oxidise into hexanal, the unpleasant-smelling fatty acid that serves as an off-putting indicator of rancidity. When Pike was asked which old food he would eat himself he said '30-year-old wheat. Baked into wholewheat bread it has practically the same sensory quality as bread made from freshly grown wheat.'

Grains are known for their longevity. Every decade uncovers more US civil defence caches of vitamin-fortified survival crackers, a legacy of the Cold War. Though they are veritable museum pieces now, in the 1970s local governments struggled to decide what to do with their vast stores of still-edible crackers. Most cities eventually tossed them out. But in March 2006 another 352,000 crackers turned up in a forgotten vault below the Brooklyn Bridge. A New York bridge inspector, sampling one, charitably described it as having 'a unique flavour'.

Vintage food discoveries may also become widely known thanks to the web. In 1997 Tony Rogers, a former employee at a Wisconsin-based chemical company, cleaned out his office desk to discover a Dolly Madison apple pie he had purchased at a gas station 8 years earlier. 'I did pick up a piece of the pie, and I actually ate some of it. It was somewhat caramelised and chewy. But … it was the same taste.' In recent years Rogers has also occasionally sampled ready meals of scalloped potatoes and ham he bought in 1994: 'They still look, smell and taste exactly as they did,' he said. 'I honestly think that if my pie was put back on a shelf in a gas station, someone would purchase it, consume it and not ask any questions.'

If they did eat it, though, wouldn't they get sick? Rogers didn't, and in the case of dried foods the answer generally seems to be no. 'When stored at room temperature in an oxygen-free atmosphere, there is no reason to believe dried food would not be safe,' said Pike. 'Microbiologically, low-moisture food that is safe when packaged is still safe after storage.' Who could disagree? Plenty of people seemed to have lived to tell the tale.

It's time to move outside the human body. They've been around for a long time. The oldest head-louse egg to be discovered was found attached to a 10,000-year-old human hair at an archaeological site in north-east Brazil, and we are still developing new ways to bump them off.

✳ Spot the louse

A shampoo developed in the US in the year 2000 provided an innovative weapon to help parents battle infestations of lice in their children's heads. The product caused lice eggs, or nits, to glow under ultraviolet light, making them much easier to spot and remove by hand.

Sydney Spiesel, a paediatrics professor at Yale University School of Medicine, formulated the shampoo after he was forced to manually remove pesticide-resistant lice and their eggs from a child's head. 'The child had thick, blonde hair. It took me an hour to go through this kid's head,' he said.

Every year about 14 million children get head lice in the US alone. For years, shampoos containing permethrin were used to keep the bugs at bay. But studies – and experience in schools and day nurseries – showed that lice had become resistant to this and other common pesticides. This left parents literally nit-picking, using special fine-toothed combs bought at pharmacies. But the tiny eggs, at less than a milli-metre across, are hard to see – and they stick tenaciously to the hair. If even a few eggs are missed, the child can quickly become re-infested.

Knowing that lice eggs contain chitin – the polysaccharide from which the exoskeletons of arthropods are made – Spiesel took a commercial organic dye he knew would bind to chitin and added it to an over-the-counter shampoo. The dye was 'delightfully cheap and delightfully non-toxic', he said. The dye bound to the chitin in the eggs, but not to the hair or scalp. When Spiesel shone a 'black' (UV) light on the eggs he saw a 'brilliant glow' which made them easy to remove.

Grossology – pretty lousy, eh?

4 Death, doctors and the human body

It's time to visit the doctor, and we hope you are paying attention. Many people don't, it seems, which leads to problems. Health research from the US reveals an unexpected medical risk factor: living too close to the doctor. The Mayo Clinic in Minnesota was worried that after an appointment many patients either forget what their doctor told them or ignored the doctor's advice anyway.

To find out, researchers monitored 556 people on visits to their GP, noted what the doctor told each patient, then followed the patient home and asked what they had heard. The patients 'did not mention 68 per cent of the health problems diagnosed,' the study found, 'including 54 per cent of the most important problems'. The study suggested that 'denial' and 'selective listening' were the main techniques used by the patients to ignore medical advice. But the survey concluded that 'patients who travelled a considerable distance for their care were most likely to remember and follow what their doctor had discussed with them'.

If you don't pay attention to the offering of extraordinary medical research contained in this chapter you'll miss out on Robert E. Cornish, the doctor who reckons he can reverse death, the researchers who made their subject swallow a balloon full of cold water as a means of cooling the human body, and what it feels like to die (once you've found out you'll need to book an appointment with Dr Cornish).

If you keep reading you'll also learn about two of this book's heroes, Stubbins Ffirth and Barry Marshall. Stubbins drank infected vomit and Barry is a Nobel prizewinner who swallowed something else.

Death also features to a worrying degree in this chapter. We have stories about execution, and what it's like to drown or bleed to death. OK, we've changed our minds: pay no attention at all to this chapter and feel a whole lot better about yourself – no visit to the doctor required.

It's always been a matter of debate whether you should read the last page of a book first, just to check that it's worth your while bothering with the earlier bits. Under normal circumstances we'd decline, but we're not talking about books here, we're talking about life. And we've decided that, in this case, the end is a very good place to start. So, to set the scene, let's establish just what life, or its absence, is exactly.

✳ When is a person dead?

In 1962, R. S. Schwa, a neurologist at Harvard University, stated that death in humans is certain if an electroencephalogram (a measure of electrical activity in the brain) showed continuous flat lines without any rhythm for an hour or more, even if the heart was beating. The electroencephalogram would also show a resistance of over 50,000 ohms and would not register a discharge even if a loud noise was made next to the patient. In addition spontaneous respiration should have ceased for an hour.

There are more sophisticated means of determining death these days, but we like the idea that you're dead if you haven't breathed for an hour and a loud noise doesn't wake you up. And when death finally happens – as it will to each and every one of us

– what does it actually feel like? Do we really know? In 2007,
New Scientist *decided to find out more.*

✳ How does it feel to die?

Is it distressing to experience consciousness slipping away or
something people can accept with equanimity? Are there any
surprises in store as our existence draws to a close? These are
questions that have plagued philosophers and scientists for
centuries, and chances are you've pondered them too
occasionally.

None of us can know the answers for sure until our own
time comes, but the few individuals who have had their
brush with death interrupted by a last-minute reprieve can
offer some intriguing insights. Advances in medical science,
too, have led to a better understanding of what goes on as the
body gives up the ghost.

Death comes in many guises, but one way or another it is
usually a lack of oxygen to the brain that delivers the coup de
grâce. Whether as a result of a heart attack, drowning or suf-
focation, for example, people ultimately die because their
neurons are deprived of oxygen, leading to cessation of electrical activity in the brain – the modern definition of biological death.

If the flow of freshly oxygenated blood to the brain is
stopped, through whatever mechanism, people tend to have
about ten seconds before losing consciousness. They may
take many more minutes to die, though, with the exact mode
of death affecting the subtleties of the final experience. Read
on for a brief guide to a few of the many and varied ways
death can suddenly strike.

Drowning

Suffocating to death in water is neither pretty nor painless, though it can be surprisingly swift. Just how fast people drown depends on several factors, including swimming ability and water temperature. In the UK, where the water is generally cold, 55 per cent of open-water drownings occur within 3 metres of safety. Two-thirds of victims are good swimmers, suggesting that people can get into difficulties within seconds, said Mike Tipton, a physiologist and expert in marine survival at the University of Portsmouth in the UK.

Typically, when a victim realises that they cannot keep their head above water they tend to panic, leading to the classic 'surface struggle'. They gasp for air at the surface and hold their breath as they bob beneath, says Tipton. Struggling to breathe, they can't call for help. Their bodies are upright, arms weakly grasping, as if trying to climb a non-existent ladder from the sea. Studies with New York lifeguards in the 1950s and 1960s found that this stage lasts just 20 to 60 seconds.

When victims eventually submerge, they hold their breath for as long as possible, typically 30 to 90 seconds. After that, they inhale some water, splutter, cough and inhale more. Water in the lungs blocks gas exchange in delicate tissues, while inhaling water also triggers the airway to seal shut – a reflex called a laryngospasm. 'There is a feeling of tearing and a burning sensation in the chest as water goes down into the airway. Then that sort of slips into a feeling of calmness and tranquillity,' said Tipton, describing reports from survivors.

That calmness represents the beginnings of the loss of consciousness from oxygen deprivation, which eventually results in the heart stopping and brain death.

Bleeding to death

The speed of exsanguination, as bleeding to death is known, depends on the source of the bleed, according to John Kortbeek at the University of Calgary in Alberta, Canada, and chair of Advanced Trauma Life Support for the American College of Surgeons. People can bleed to death in seconds if the aorta, the major blood vessel leading from the heart, is completely severed, for example, after a severe fall or car accident.

Death could creep up much more slowly if a smaller vein or artery is nicked – even taking hours. Such victims would experience several stages of haemorrhagic shock. The average adult has 5 litres of blood. Losses of around 750 millilitres generally cause few symptoms. Anyone losing 1.5 litres feels weak, thirsty and anxious, and would be breathing fast. By 2 litres, people experience dizziness, confusion and then eventual unconsciousness.

'Survivors of haemorrhagic shock describe many different experiences, ranging from fear to relative calm,' Kortbeek said. 'In large part this would depend on what and how extensive the associated injuries were. A single penetrating wound to the femoral artery in the leg might be less painful than multiple fractures sustained in a motor vehicle crash.'

Fire
Long the fate of witches and heretics, burning to death is torture. Hot smoke and flames singe eyebrows and hair and burn the throat and airways, making it hard to breathe. Burns inflict immediate and intense pain through stimulation of the nociceptors – the pain nerves in the skin. To make matters worse, burns also trigger a rapid inflammatory response, which boosts sensitivity to pain in the injured tissues and surrounding areas.

As burn intensities progress, some feeling is lost but not much, according to David Herndon, a burns-care specialist at University of Texas Medical Branch in Galveston. 'Third-degree burns do not hurt as much as second-degree wounds, as superficial nerves are destroyed. But the difference is semantic; large burns are horrifically painful in any instance.'

Some victims of severe burns report not feeling their injuries while they are still in danger or intent on saving others. Once the adrenalin and shock wear off, however, the pain quickly sets in. Pain management remains one of the most challenging medical problems in the care of burns victims.

Most people who die in fires do not in fact die from burns. The most common cause of death is inhaling toxic gases – carbon monoxide, carbon dioxide and even hydrogen cyanide – together with the suffocating lack of oxygen. One study of fire deaths in Norway from 1996 found that almost 75 per cent of the 286 people autopsied had died from carbon monoxide poisoning.

Depending on the size of the fire and how close you are to it, concentrations of carbon monoxide could start to cause headache and drowsiness in minutes, eventually leading to unconsciousness. According to the US National Fire Protection Association, 40 per cent of the victims of fatal home fires are knocked out by fumes before they can even wake up.

Fall from a height

A high fall is certainly among the speediest ways to die: terminal velocity (no pun intended) is about 200 kilometres per hour, achieved from a height of about 145 metres or more. A study of deadly falls in Hamburg, Germany, found that 75 per cent of victims died in the first few seconds or minutes after landing.

The exact cause of death varies, depending on the landing surface and the person's posture. People are especially unlikely to arrive at the hospital alive if they land on their head – more common for shorter (under 10 metres) and higher (over 25 metres) falls. A 1981 analysis of 100 suicidal jumps from the Golden Gate Bridge in San Francisco – height: 75 metres, velocity on impact with the water: 120 kilometres per hour – found numerous causes of instantaneous death including massive lung bruising, collapsed lungs, exploded hearts or damage to major blood vessels and lungs through broken ribs.

Survivors of great falls often report the sensation of time slowing down. The natural reaction is to struggle to maintain a feet-first landing, resulting in fractures to the leg bones, lower spinal column and life-threatening broken pelvises. The impact travelling up through the body can also burst the aorta and heart chambers. Yet this is probably still the safest way to land, despite the force being concentrated in a small area: the feet and legs form a 'crumple zone' which provides some protection to the major internal organs.

Some experienced climbers or skydivers who have survived a fall report feeling focused, alert and driven to ensure they land in the best way possible: relaxed, legs bent and, where possible, ready to roll. Certainly every little helps, but the top tip for fallers must be to aim for a soft landing. A paper from 1942 reported a woman falling 28 metres from her apartment building into freshly tilled soil. She walked away with just a fractured rib and broken wrist.

Explosive decompression

Death due to exposure to vacuum is a staple of science fiction plots, whether the unfortunate gets thrown from an airlock or ruptures their spacesuit. In real life there has been just one fatal space depressurisation accident. This occurred on the Russian Soyuz-11 mission in 1971, when a seal leaked upon re-entry into the Earth's atmosphere; upon landing all three flight crew were found dead from asphyxiation.

When the external air pressure suddenly drops, the air in the lungs expands, tearing the fragile gas exchange tissues. This is especially damaging if the victim neglects to exhale prior to decompression or tries to hold their breath. Oxygen begins to escape from the blood and lungs. Up to 40 seconds after the pressure drops, bodies begin to swell as the water in tissues vaporises, though the tight seal of skin prevents them from 'bursting'. The heart rate rises initially, then plummets.

Bubbles of water vapour form in the blood and travel through the circulatory system, obstructing blood flow. After about a minute, blood effectively stops circulating.

Human survivors of rapid decompression accidents include pilots whose planes lost pressure, or in one case a NASA technician who accidentally depressurised his flight suit inside a vacuum chamber. They often report an initial pain, like being hit in the chest, and may remember feeling air escape from their lungs and the inability to inhale. Time to loss of consciousness was generally less than 15 seconds.

Surprisingly, in view of these apparently traumatic effects, animals that have been repressurised within 90 seconds have generally survived with no lasting damage.

Obviously some of the above must be hearsay, until it happens to us we won't know – and we're in no rush to find out. Some unfortunates, however, do find out before their time would otherwise be up, thanks to the human practice of execution. And other unfortunates have been, well, unfortunate enough to study them. In 1983, one brave New Scientist *correspondent did just that.*

✳ An unnatural way to die

In 1983, the British parliament voted decisively against the reintroduction of the death penalty. But what were they voting against? What happens during an execution? Why does the person die? Is death instantaneous? Which is the most humane method?

If the death penalty had been reintroduced in the UK, hanging would probably have been the method. The blindfolded prisoner stands on a trapdoor with a rope around his or her neck. The door is opened suddenly and the weight of the prisoner's falling body causes traction and tearing of the cervical muscles, skin and blood vessels. The upper cervical vertebrae are dislocated, and the spinal cord is separated from the brain: this is the lesion which causes death. The volume of blood in the skull and face quickly increases, but soon falls again. The respiratory and heart rates slow until they stop and death supervenes.

Initially during hanging the prisoner attempts to move, presumably reacting mainly to the pain of neck traction and dislocation. Later there is a series of reflex movements, as a result of spinal reflexes originating at the site of severance of the brain from the spinal cord, but these are not evidence that he or she can still feel. Hanging, however, does not immediately arrest respiration and heartbeat. They both start to slow immediately, but whereas breathing stops in seconds, the heart may beat for minutes. Blood loss plays little part in death due to hanging.

It is impossible to know for how long the condemned person feels pain, but in addition to cervical pain, the prisoner probably has an acute headache from the occlusion of veins and engorgement of cerebral blood vessels; this results from the rope closing off the veins of the neck before occluding the carotid and vertebral arteries. In experiments during the Second World War on human volunteers, the pressure at the lower end of the neck was raised to 600 millimetres of mercury and consciousness was lost in 6 to 7 seconds. This would have been sufficient time to feel pain.

Shooting is probably the second most widely used execution technique. Death is virtually instantaneous if the person is shot at close quarters through the skull; the bullet penetrates the medulla, which contains the vital respiratory and cardiac centres, among others.

But condemned prisoners are usually shot by firing squads aiming at the heart from some metres away and it is difficult to shoot a person dead with a single or few shots, except at very close range. The reason for this is that the cause of death in these cases is normally blood loss through rupture of the heart or a large blood vessel, or tearing of the lungs. Any lover of opera or westerns knows that death in these circumstances takes several minutes, quite enough for the victim to sing a powerful aria or almost describe the location of buried treasure.

Bullets, of course, cause a great deal of damage in tissues. High-velocity missiles, such as rifle bullets, have a tremendous amount of energy, which is partly released as heat within the tissues. This causes evaporation of tissues and water, forming a carrot-shaped space several hundred times the volume of the original bullet. When the bullet has passed through, this cavity collapses and sucks in dead tissue and contaminated air.

The guillotine was named after the French deputy who proposed its use in 1789 and introduced as a swift and painless device to extend to all citizens the advantages of a technique used only on noblemen in France. It was considered more humane because the blade was sharper and execution was more rapid than was normally accomplished with an axe. Death occurs due to separation of the brain and spinal cord, after transection of the surrounding tissues. This must cause acute and possibly severe pain. Consciousness is probably lost within 2 to 3 seconds, due to a rapid fall of intracranial perfusion of blood.

Some macabre historical reports from post-revolutionary France cited movements of the eyes and mouth for 15 to 30 seconds after the blade struck, although these may have been post-mortem twitches and reflexes.

Garrotting was used in the Iberian peninsula until the 1970s. It is a form of strangulation by a metal collar with a clamp. Those who use it believe that the resultant dislocation of the neck is rapid and death is instantaneous. Unfortunately, although the clamp is tightened quickly, the degree of compression of the neck sufficient to dislocate it takes some seconds to achieve. The neck tissues are tough and the application of the contraption is highly disagreeable. In addition to compressing the soft tissues, the clamp occludes the trachea. Therefore it kills by asphyxia, cerebral ischaemia and neck dislocation. Dying is painful, deeply distressing and may take several minutes.

Electrocution was first approved by the state of New York in 1888, and several hundred people a year were executed in the US for rape and armed robbery, as well as murder, until the end of the 1960s.

The prisoner is fastened to a chair by his chest, groin, arms and legs to prevent violent movements and to keep the electrodes in place. These are moistened copper terminals attached to one calf and a band round the head. Jolts of 4 to 8 amperes at levels between 500 and 2000 volts are applied for half a minute at a time, and a doctor inspects the condemned to decide if death has occurred or another jolt should be administered.

In 1982, John Louis Evans was shocked for half a minute. This broke the leg electrode, which was reattached. A second attempt failed to kill him and smoke was seen coming out of his mouth and left leg. He was given a third dose. It took ten minutes before the attending physician certified him as being dead.

The effects of accidental electrocution are burns, respiratory paralysis and cardiac arrest. The electric chair was introduced because it was believed it causes instant and painless death. Close observation has shown this is clearly not the case. There is no reason whatsoever to believe that the condemned person does not suffer severe and prolonged pain. The prisoner is so firmly fastened to the chair that he cannot move. The large amount of energy in the shock paralyses the muscles, presumably leading to the belief that the failure to move meant the prisoner was not suffering pain. However, a prisoner being electrocuted is paralysed and asphyxiated, but almost certainly fully conscious and sentient with a feeling of being burnt to death while conscious of the inability to breathe. It must feel very similar to the medieval trial by ordeal of being dropped into burning oil.

Two further techniques have been introduced in the US, examples of which are given below. The first is intravenous injection. In December 1982, Charlie Brooks of Texas had his vein cannulated by a physician. Then, from outside the execution chamber and unseen by the prisoner, a mixture was injected: this consisted of the rapidly acting anaesthetic Pentothal-curare to paralyse the muscles, and potassium chloride to stop the heart. The condemned man would have gone off to sleep in 10 to 15 seconds, never to wake again. The prisoner would have suffered no more pain than a patient being given Pentothal before an operation. However, the physician gave an overdose and the condemned man died under anaesthetic from central respiratory depression.

Other states have chosen gassing. In September 1983, Jimmy Lee Gray was strapped to a chair in an airtight room. Sodium cyanide crystals were dropped into a bath of sulphuric acid below his chair by depressing a lever from outside. This created hydrogen cyanide gas and the condemned man inhaled it. A sufficient concentration to constitute a lethal dose would take several seconds to accumulate, depending on how hard he tried to avoid inhaling. It would cause acute difficulty in breathing, asphyxia and, possibly, pain in the stomach. The prisoner would be severely distressed and in pain during the whole procedure. The resulting hypoxia would cause him to have spasms as in an epileptic fit, visible if he were not bound firmly. The prisoner would have died of inhibition of respiratory enzymes.

This article does not discuss the morality of capital punishment. A physician has no more expertise in general moral and ethical questions than does anyone else. These are simply the physiological facts of execution. They are inescapable if gruesome, but, if considered, may well help us to make appropriate moral decisions.

Nasty stuff indeed – and enough to make us determined prohibitionists on a subject that is still being debated today. To be honest though, given the choice, we'd go for the guillotine. Got to be better than an old axe, for sure. Apparently it took the axeman three attempts to sever the head of Mary Queen of Scots in 1587. He had to finish the job with a knife. Decades earlier in 1541, Margaret Pole, the Countess of Salisbury, was being executed at the Tower of London. She was dragged to the block but refused to lay her head down. The inexperienced axeman made a gash in her shoulder rather than her neck. According to some reports, she leapt from the block and was chased by the executioner, who struck 11 times before she died. Lovely. But although we now know what it's like to die, what really does happen afterwards?

⚛ What happens after you die?

Many have tried to harness the rational horsepower of science to answer this most floaty question. Some were physicians, some physicists, some psychologists. Two were Nobel prize-winners. One was a sheep rancher. They have tackled it in labs, in hospital operating rooms, in barns behind their houses. Of them, only one landed an irrefutable proof – not a suggestive nugget or an inexplicable anomaly, but the sort of answer you could plant your flag into and say, 'Victory! Now I know for certain.' The man's name was Thomas Lynn Bradford.

Though his background was in electrical engineering, Bradford's afterlife experiment involved gas, not electricity. On 6 February 1921, Bradford sealed the doors and windows of his rented room in Detroit, Michigan, blew out the pilot on his heater, and turned on the gas.

Finding out is easy. Reporting back is the challenge. For this Bradford needed an accomplice. Some weeks back, he had placed a newspaper advertisement seeking a fellow spiritualist to help him with his quest. One Ruth Doran responded. The two met and agreed, as the *New York Times* put it, 'that there was but one way to solve the mystery – two minds properly attuned, one of which must shed its earthly mantle'. The protocol was sloppy at best, for regardless of whether or not our mantle-shucking engineer came through on the telepathic wireless, Mrs Doran, for the sake of spiritualism or publicity, could simply have told the reporters that he did. But she did not lie. *The Times* ran a follow-up under the headline 'Dead Spiritualist Silent'.

A better-pedigreed variation of the Bradford experiment was undertaken by the physicist Oliver Lodge, once the principal of the University of Birmingham. Prior to his death in 1940, he devised the Oliver Lodge Posthumous Test. The goal, again, was to prove the existence of life after death. Lodge composed a secret message and sealed it in a packet (the Oliver Lodge Posthumous Packet) so that when, after his death, he told mediums (four of them, recruited by the Oliver Lodge Posthumous Test Committee) what the message was, their stories could be checked.

The packet itself was sealed inside seven envelopes, each envelope containing a clue the mediums could employ to jog the deceased physicist's memory should he forget his own secret. Instead, the clues merely irritated the mediums. The contents of Envelope 3, for example, read: 'If I give a number of 5 digits it may be correct, but I may say something about 2 8 0 1, and that will mean I am on the scent. It is not the real number … but it has some connection with it. In fact it is a factor of it.' Eventually the mediums walked off the set and the Posthumous Packet was torn open, leaving the committee with nothing for their efforts but a slip of paper bearing an obscure musical fragment and a gnawing suspicion that Sir Oliver had been a few envelopes short of a stationery set. Of course, even had the mediums succeeded, one could never have been certain whether they might simply have – via

some discreet Oliver Lodge Posthumous Envelope Steaming
– peeked.

*The dead-researcher approach is clearly not the way to go. A more
promising tack might be to focus on those who have not quite
died, but merely managed a sneak preview – in the form of a near-
death experience.*

✳ Out-of-body experiences (and weighing the soul)

If someone could prove that a near-death experience is, in
verifiable fact, a round-trip visit to some other dimension and
not a mirage of the dying mind, that would surely be something to hang one's hopes on. But how does the person who
claims to have glimpsed the beyond go about proving it?
There seems to be no afterlife gift shop, no snow globes full
of angel dandruff. Best to focus on one of those near-death
trips that take the traveller only as far as the ceiling, enabling
a reconnaissance-type view of one's corporeal hull down
below. If one could at least prove that one had seen the details
of the room from up there – and not remembered or halluci-
nated or some combination of the two – then that would at
least establish the possibility of the seeming impossibility of
a consciousness existing independent of its biological
moorings.

And that is why a laptop computer was duct-taped to the
highest monitor in a cardiology operating room at the University of Virginia in Charlottesville. The computer had been
programmed to show, for the duration of each operation, one
of 12 images, chosen at random and unknown to anyone,
including the researchers. The laptop would be flat open with
the screen facing the ceiling, such that the only way a surgery

patient might view the image would be as a disembodied consciousness. As patients came out of anaesthesia, psychologist Bruce Greyson interviewed them about what they remembered of their time in the operating room. So far there have been no surprises. Other, that is, than the surprising cooperation of a team of cardiac surgeons. Heart surgeons who believe that a consciousness can occasionally perceive things in an extrasensory manner, independent of a brain and eyeballs, are less rare than you might think.

But even then, how would we know that the near-death experience isn't a hallmark of dying, not death – a stopover, not a final destination? How do we know that several minutes later the bright light doesn't dim and the euphoria fade and you're just flat-out non-existent? 'We don't know,' concedes Greyson. 'It's possible it's like going to the Paris airport and thinking you've seen France.'

Another way to approach the afterlife would be to consider the vehicle: the soul (or consciousness, if you like). If the soul were something you could weigh, like a pancreas or a wart, then proving that it abandons the corpse at death would be a simple matter of placing a dying person on a scale and watching to see if the needle went down at the moment he died (while also accounting for the minute amount of weight lost via moisture in exhalations and sweat).

This is exactly what a Massachusetts physician named Duncan MacDougall did, beginning in 1901, using a tricked-out industrial silk scale (see page 43). His post at a tuberculosis sanatorium provided MacDougall with a steady source of study subjects. He weighed six men as they died, and there was, he said in a series of articles in *American Medicine*, always a down-tick of the needle. However, only one of his trials went off without a significant hitch. Twice the authorities barged into the room and tried to stop the proceedings. Oafish accomplices jostled the scale. Subjects died as the scale was being zeroed. And so MacDougall's claimed proof – that the soul exists, and that it weighs about 20 grams – is really no more than anecdote.

⚛ Sheep souls and thermodynamics

Some 90 years later, a sheep rancher in Bend, Oregon, tried to replicate MacDougall's work. When a local hospital rebuffed his solicitation for terminal patients, Lewis Hollander Jr turned to his flock. Interestingly, he found that sheep momentarily gain a small amount of weight at the moment they die, suggesting that the answer to the question 'What happens when we die?' might in fact be: 'Our souls go into sheep.'

Of course, it's a stretch to think that the weight of a soul would register on a scale built for the likes of livestock or bolts of cloth. But what if you were to get your hands on a scale calibrated not in ounces or grams, but in picograms – trillionths of a gram? If you consider consciousness to be information energy, as some do, then it would have a (very, very, very teeny tiny) mass. And if you were to build a closed system, such that no known sources of energy could leave or enter undetected, and you rigged it up to your picogram scale, and put a dying organism inside this system, then you could, in theory, do the MacDougall.

In 2006, Duke University professor Gerry Nahum was very keen to undertake a consciousness-weighing project of his own (using not sheep nor men but leeches). Though he taught gynaecology and obstetrics, Nahum had a background in thermodynamics and information theory and had even worked out a 25-page proposal of exactly how to do it, if only someone would fund him the $100,000 he estimated it would cost.

If consciousness is energy, then you probably don't need proof that it survives death, because proof already exists: the First Law of Thermodynamics – energy is neither created nor destroyed. Though it's hard to take much comfort from this. Who wants to spend eternity as a blip, a gnat's fart, of disordered energy, with no brain at their disposal to help them remember or imagine or solve the Sunday crossword? What would it be like? Would there even be a be? Nahum used the analogy of the computer: perhaps you'd be the operating system, stripped of its programs and interfaces. Heaven as the back of the closet where the broken-down Dells and Compaqs go.

If we are to eventually have our answer, our proof, it will no doubt come to us courtesy of quantum theory, or whatever takes its place. Few of us will understand it well enough to take much comfort, however, if indeed comfort is what it offers. Try to enjoy life without worrying about the 'after' bit, and keep in mind that one day altogether too soon, bad luck or genetics will hand you the answer. In the meantime, be nice to sheep.

Cripes. Who knows what's going on? Perhaps we'd be better off visiting Dr Cornish.

⚛ Reversing death

Robert E. Cornish, a researcher at the Berkeley campus of the University of California during the 1930s, believed he had found a way to restore life to the dead – at least in cases where major organ damage was not involved. His technique involved seesawing corpses up and down to circulate the blood while injecting a mixture of adrenalin and anticoagulants. He tested his method on a series of fox terriers, all of whom he named Lazarus after the biblical character brought back to life by Jesus.

First Cornish asphyxiated the dogs and left them dead for ten minutes. Then he attempted to revive them. His first two trials failed, but numbers three and four were a success. With a whine and a feeble bark, the dogs stirred back to life. Though blind and severely brain damaged, they lived on for months as pets in his home, reportedly inspiring terror in other dogs.

Cornish's research provoked such controversy that the University of California eventually ordered him off the campus. He continued his work in a tin shack attached to his house, despite complaints from neighbours that mystery fumes from his experiments were causing the paint on their homes to peel.

Many years later, in 1947, Cornish announced he was ready to experiment on a human being. He now had a new tool in his arsenal: a home-made heart-lung machine built out of a vacuum cleaner blower, radiator tubing, an iron wheel, rollers and 60,000 shoelace eyes. Thomas McMonigle, a prisoner awaiting execution on death row, volunteered to be his guinea pig, and Cornish asked the state of California for permission to proceed with his experiment. After some deliberation, the state turned him down. Apparently officials were worried that, should McMonigle come back to life, they might have to free him.

Disheartened, Cornish retreated to his home, where he eked out a living selling a toothpaste of his own invention.

We find it most surprising that he couldn't find anyone other than a death-row prisoner to be a volunteer. That aside, we're not sure why Cornish didn't attempt the process on himself if he was so certain – perhaps he didn't have the confidence of the brave researchers who follow …

✵ Hard to swallow

As a junior doctor at the Royal Perth Hospital, Western Australia, Barry Marshall was so sure the medical establishment was wrong about the cause of stomach ulcers that he swallowed the bacteria he believed were to blame. It still took years to convince everyone – but it was to win him a share in a Nobel prize, alongside Robin Warren. *New Scientist* interviewed him in 2006.

What made you decide to swallow the bacteria?
It was so frustrating to see ulcer patients having surgery, or even dying, when I knew a simple antibiotic treatment could fix the problem. Back in 1984, conventional medical wisdom was that ulcers were caused by stress, bad diet, smoking, alcohol and susceptible genes – and that no bacteria could survive in the stomach. Working with pathologist Robin Warren, I found a bacterium called *Helicobacter pylori* in all duodenal ulcer patients and in 77 per cent of those with gastric ulcers. We tried to infect animals to prove this bacterium was the culprit but that failed, so we had to find a human volunteer.

Why did it have to be you?
I was the only person informed enough to consent, so I decided to be my own guinea pig. I didn't seek approval from the hospital's ethics committee because I didn't want to risk being turned down, and I didn't even tell my wife until I had swallowed the bacteria. By then I had successfully treated several patients suffering from *H. pylori* infections using antibiotics, so it seemed that I had a cure.

What happened after you swallowed the bacteria?
I was fine for three days, then began to feel nauseous, and soon began vomiting. My wife told me I had 'putrid breath'.

After ten days, a biopsy confirmed the bacteria had infected my stomach, and the stomach wall was inflamed with gastritis, which can eventually lead to ulcers. My experiment overturned 100 years of knowledge about ulcers. We published our results in *The Lancet*.

Was it hard to convince the world that such a 'miracle cure' existed?
Yes. When *The Lancet* finally used the word 'cure' in 1989 we thought that everybody must believe us now, but it was another 8 long years before most people in western countries were aware that *H. pylori* caused ulcers. In the meantime, millions of people had taken unnecessary drugs or had surgery, at a cost of billions of dollars.

Did you find that infuriating?
At the time I thought it was somewhat immoral, because doctors who were sceptical about *H. pylori* were making decisions that permanently affected the lives of their patients. It was very easy for them to stick with the old treatments. I was annoyed about the level of opposition to our theory, and that people were not testing it, but now I realise that it takes time for an idea to gain acceptance.

And Barry Marshall wasn't the only one prepared to swallow something horrid.

⚛ The vomit-drinking doctor

How far would you go to prove your point? Stubbins Ffirth, a doctor-in-training living in Philadelphia during the early 19th century, went further than most. Way further.

Having observed that yellow fever ran riot during the summer, but disappeared over the winter, Ffirth hypothesised it was not a contagious disease. He reckoned it was caused by an excess of stimulants such as heat, food and noise. To prove his hunch, Ffirth set out to demonstrate that no matter how much he exposed himself to yellow fever, he wouldn't catch it.

He started by making a small incision in his arm and pouring 'fresh black vomit' obtained from a yellow-fever patient into the cut. He didn't get sick.

But he didn't stop there. His experiments grew progressively bolder. He made deeper incisions in his arms into which he poured black vomit. He dribbled the stuff in his eyes. He filled a room with heated 'regurgitation vapours' – a vomit sauna – and remained there for two hours, breathing in the air. He experienced a 'great pain in my head, some nausea, and perspired very freely', but was otherwise OK.

Next Ffirth began ingesting the vomit. He fashioned some of the black matter into pills and swallowed them down. He mixed half an ounce of fresh vomit with water and drank it. 'The taste was very slightly acid,' he wrote. 'It is probable that if I had not, previous to the two last experiments, accustomed myself to tasting and smelling it, that emesis would have been the consequence.' Finally, he gathered his courage and quaffed pure, undiluted black vomit fresh from a patient's mouth. Still he didn't get sick.

Ffirth rounded out his experiment by liberally smearing himself with other yellow-fever tainted fluids: blood, saliva, perspiration and urine. Healthy as ever, he declared his hypothesis proven in his 1804 thesis.

He was wrong. Yellow fever, as we now know, is very contagious, but it requires direct transmission into the bloodstream, usually by a mosquito, to cause infection.

Considering the strenuous efforts Ffirth took to infect himself, it must be considered something of a miracle he remained alive. The bright spot for him was that, after all he put himself through, the University of Pennsylvania did award him the degree of Doctor of Medicine. What his patients made of him unfortunately remains unrecorded.

It's a cliché, but some people really are prepared to put their bodies on the line.

⚛ This won't hurt a bit

Two surgeons had finished work for the day. But instead of going home, they began to prepare for one more operation – a little out-of-hours experiment intended to advance the art of anaesthesia. August Bier was a rising star at the Royal Surgical Clinic in Kiel, north Germany. His young assistant, August Hildebrandt, had agreed to help him.

What happened next was not so much heroic as comic. Just one little mistake and courageous selflessness turned to black comedy. It made Bier's name. But the events of that evening would be forever etched on Hildebrandt's memory, not to mention several other parts of his body.

In the 1890s, general anaesthesia was decidedly dodgy. Chloroform sent patients gently to sleep – but there was no room for mistakes. A few drops too many and the patient would be dead before the surgeon picked up the scalpel. Ether wasn't quite so dangerous, but it was slow to act – surgeons sometimes started to operate before their patients had gone under. The survivors suffered unpleasant side effects – from violent headaches and vomiting to ether pneumonia.

Bier reasoned it should be possible to banish sensation from most of the body without knocking the patient out completely by injecting a small dose of cocaine into the cerebrospinal fluid that bathes the spinal cord. He tried his technique on half a dozen patients. They lost sensation from the lower part of their bodies long enough for him to carve out chunks of diseased bone from their ankles, knees and shins – and even the thigh and pelvis. 'On the other hand, so many complaints had arisen in association with this method that they equalled the complaints usually occurring after general anaesthesia,' he wrote. 'To arrive at a valid opinion, I decided to conduct an experiment on my own body.'

The procedure was simple enough. Hildebrandt had to make a lumbar puncture by plunging a large needle through the membranes that protected Bier's spinal cord into the fluid-filled space beneath. Then he had to fit a syringe on the needle and inject a solution of cocaine. But preparations for the experiment had been less than meticulous.

Hildebrandt made the lumbar puncture. Then, with his finger over the hub of the needle to prevent fluid from leaking out, he took up the syringe of cocaine – only to find it was the wrong fit. As he fumbled with the needles, Bier's cerebrospinal fluid began to squirt out. Horrified, Hildebrandt stopped and plugged the wound. This was when the pair should have called it a day. Instead, Hildebrandt offered to take Bier's place.

At 7.38 pm, after checking the needles more carefully, Bier began. The cocaine worked fast. 'After 7 minutes: Needle pricks in the thigh were felt as pressure; tickling of the soles of the feet was hardly felt.' Bier jabbed Hildebrandt in the thigh with a needle. Nothing. He tried harder, stabbing the thigh with the surgical equivalent of a stiletto. Still no response. Then, 13 minutes into the experiment, Bier stubbed out a cigar on Hildebrandt's leg.

Bier now wanted to know how far the insensitivity extended, and invented a simple test. 'Pulling out pubic hairs was felt in the form of elevation of a skinfold; pulling of chest hair above the nipples caused vivid pain.' So now he knew. It was more than 20 minutes since Hildebrandt had stopped feeling pain. How much more could he take? Bier increased his efforts. He smashed a heavy iron hammer into

Hildebrandt's shin bone and then, when that failed to have any effect, gave his testicles a sharp tug. In a final burst of enthusiasm, Bier stabbed the thigh right to the bone, squashed hard on a testicle and, for good measure, rained blows on Hildebrandt's shin with his knuckles.

After 45 minutes, the effect of the cocaine began to wear off. The two surgeons, one missing a significant amount of cerebrospinal fluid, the other battered, burnt and suffering from serious stab wounds, went out for dinner. 'We drank wine and smoked several cigars,' wrote Bier.

The next morning, Bier woke feeling bright and breezy. By the afternoon he had turned pale, his pulse was weak and he felt dizzy whenever he stood up. 'All these symptoms disappeared as soon as I lay down horizontally, but they returned when I arose. In the late afternoon, therefore, I had to go to bed.' He stayed there for the next 9 days. When he finally got up again he felt quite well. 'I was perfectly able to tolerate the strain of a week's hunting in the mountains,' he wrote.

Hildebrandt didn't escape so lightly. The first night he was violently ill. He had a splitting headache and was sick. But someone had to tend to the clinic's patients and, with Bier in bed, the job fell to him. Each morning for the next week, Hildebrandt dragged himself to work. Each afternoon, he staggered home and collapsed into bed. 'Dr Hildebrandt's legs were painful, and bruises appeared in several places,' wrote Bier, rather understating the case.

When Bier wrote his ground-breaking paper describing the experiment, he gave a blow-by-blow account of what Hildebrandt had endured. As far as Bier was concerned, the experiment was a huge success. He had shown that a tiny dose of cocaine could deaden sensation for long enough to perform a major operation. Spinal anaesthesia was far safer than general anaesthesia, and within two years surgeons around the world were using it. Bier put the headaches down to the loss of cerebrospinal fluid, and he was right – this was finally proved in the 1950s.

Hildebrandt, though, had gone right off Bier and became one of his most vehement critics. When a row blew up over who had really been first to invent spinal anaesthesia, Hildebrandt championed Bier's rival, an American neurologist called James Corning. Hildebrandt never said why. Perhaps he was shocked by the zeal with which Bier had battered him. Maybe he was miffed because in the end Bier was recognised as a pioneering surgeon, while he was forever known as the man whose boss had tugged his testicles.

Some researchers, of course, prefer to put other people's bodies on the line ...

❄ The fall guy

Mark Grabiner spent the best part of 15 years tripping people up. It wasn't that he was a prankster. He'd simply been trying to solve a problem that all of us face from time to time: how to avoid falling when you trip.

It has been known for decades that older people fall more than younger ones, and falls are responsible for the majority of injuries to people over the age of 65. But to understand why, Grabiner reasoned, why not study those people who usually manage to avoid it: the youth? In 1986 Grabiner decided to study young, healthy people's responses to being tripped. But first, he had a problem. If volunteers knew they were going to be tripped – and they had to so they could give their permission – how could he surprise them? Grabiner, at the University of Illinois at Chicago, worked out a way to do it.

For his experiments, he strapped the volunteers into a harness attached to a sliding track mounted in the ceiling. This caught them just before they hit the floor. 'They have to be able to throw themselves as hard as they can at the ground – and miss,' he said. Once the hapless subjects got used to their new outfits, the researchers warned them they were going to be tripped. But they didn't say when or how. 'We have to be pretty devious,' he said.

An assistant uncoiled a rope, laid it across the lab and instructed the volunteers to walk over it normally. The victims approached the rope cautiously, but relaxed once they had passed it, thinking they were safe. 'And then we trip them,' Grabiner said. An aluminium box popped out of the floor, catching the unsuspecting victim's foot as they swung it forward.

This work produced puzzling results. When people trip, the upper body pitches forward. Because older people have weaker muscles, they are less stable when they walk. So Grabiner expected to find that weak people would pitch further forward than stronger people. But this was not what his results showed. Weak or strong, people always tilted to about the same angle when they were tripped. He also found it as difficult to get strong young people to fall down as their weaker cohorts.

Unlike the younger participants, most of the elderly volunteers were easily floored. But there was a complication. The study found there were two categories of older faller: the slow and the fast. The first group knew they were weak, and tended to shuffle along slowly. In this way they gave themselves extra time to react after they tripped – though not enough, it turned out, to prevent a fall. More unexpected were the fast fallers. These were elderly people with strong back muscles and more confidence. They walked as quickly as young people, but were far easier to bring down. Why should this be?

The answer eventually came from Mirjam Pijnappels at the Free University, Amsterdam, in the Netherlands. And there seemed to be more to it than brute strength. Pijnappels improved on Grabiner's set-up by hiding not one but 21 pop-up obstacles in the floor, making it easier to trip people up by surprising them more. She also measured the distance and height that her victims moved when they were tripped. From this, she found that people used the leg that had not been tripped to win them some extra time to recover. When one leg was tripped, they immediately pushed off with the other leg, as if hopping, but also bending their knee. This pushed them 40 per cent higher into the air than in a normal step, and gave them 63 per cent more time in the air before they landed again.

According to James Ashton-Miller at the University of Michigan, these studies pointed to one crucial factor: the importance of muscle power, particularly in the calf. By power he meant not merely the force a muscle can exert, but also how fast it can produce that force. Even allowing for differences in reaction time, calf muscles in elderly people take twice as long as to reach their maximum force as young people's muscles do, which prevents them using their supporting leg in the same way.

And this observation could be put to practical use. 'The good news is that at any age muscle is trainable,' Ashton-Miller said. For older people in good health, the potential fast fallers, exercises like dancing and skipping can increase muscle power. Staying active should make it easier to push off on that supporting leg to recover from a trip. However, there was another important risk factor. In the labs, people who weren't paying attention fell more often than people who were. You may have the agility of a spring lamb, but it won't help you if you are too drunk or too engrossed in your mobile phone to react before you fall on your face.

Tripping people up, however, seems quite tame when you consider what they were up to in the 1960s.

❊ Effects of hallucinogenic drugs in blind people

How do drugs such as LSD and mescaline produce their unusual psychedelic effects and visual hallucinations? These consist of brightly coloured lights, geometric designs and impairment of colour values and space perception. In 1961, Dr Adrian M. Ostfeld of the University of Illinois showed that hallucinogenic drugs produce abnormal changes in the retina, affecting the functions of the rod-cell light receptors of the retina, probably by interfering with their oxygen supply.

He then gave LSD to 18 blind people who had not been blind from birth. A number reported 'seeing' white lights and coloured flashes, similar to those described by sighted patients who had the visual area in the occipital part of their brain stimulated electrically under local anaesthesia. But when people who had been blind since birth were given LSD they did not experience these visual hallucinations, presumably because they had no appreciation of light or colour. They did, however, experience auditory, olfactory and tactile hallucinations. One made the comment that after taking LSD the Braille print he was reading seemed to jump off the page when he touched it.

We wonder if the volunteers were asked to sign a waiver? It seems difficult to imagine a similar experiment getting the go-ahead today. And pity the poor volunteer having to swallow this.

❊ Keeping cool

In 1957, an experiment requiring a patient to swallow a balloon which was subsequently inflated in his stomach with cold water was carried out by experts who were studying ways of cooling the body sufficiently to allow operations within a dry heart. By lowering the body temperature, the oxygen needs of the tissues were reduced, and the veins leading to the heart could then be safely clamped for a short time, cutting off circulation.

The balloon of cold water worked by cooling a large volume of blood vessels in the abdomen. Recovery could be aided by circulating warm water through the balloon. Patients could not be cooled below 28 °C and the period of circulatory arrest could not exceed 10 minutes, which left about 8 minutes for running repairs inside the heart.

At a maximum of 8 minutes we can only presume that quite often one swallow doth not a recovery make. Now on to matters of great medical import. Tickling for instance ...

✳ It takes two

Tickling is a serious business. In the early 1970s, psychologists from both Oxford and Sheffield tried to find out just why tickling oneself isn't funny. In a paper published in the journal *Nature,* they suggested that it wasn't quite as simple as one might have expected.

Charles Darwin was intrigued by the ineffectiveness of the self-administered tickle, but his interest was mainly in the biological value of ticklishness and laughter. The authors of the *Nature* paper (L. Weiskrantz and C. Darlington of Oxford, and J. Elliott of Sheffield), on the other hand, set out to analyse what mechanisms are actually involved. To do this they lined up 30 intrepid volunteers brave enough to be stimulated by a Heath Robinson-type tickle apparatus.

Three types of tickle were administered: the first by an experimenter, the tickled subject remaining passive; the next was self-administered by the subject who moved the handle of the tickle apparatus; the third was done by an experimenter, but with the tickled subject's hand on the handle. The idea of this last method was to see if there was any effect on the ticklishness when the subject went through the motion of tickling but was not actually in command.

Sure enough, the efficacy of tickling was greatest when done by someone else, least when self-administered, and intermediate when the subject participated in arm movements. Clearly, this participation helped eliminate ticklishness, but not completely. Weiskrantz and his colleagues thought there might be two factors at play in self-tickling. One was a sort of negative feedback from the arm movements doing the tickling (or going through the motions, as in the experiment). The other was the 'command signal' generated when the subject decided to tickle himself.

❋ Spectacularly smart

It was a once-prevalent myth that if you wore spectacles when you were 8 years old, you were going to develop phenomenal brainpower. No substantiation of the notion was ever forthcoming, yet an experiment in 1971 by Michael Argyle and R. McHenry of the Institute of Experimental Psychology, Oxford, confirmed the fact that many people still felt spectacle wearers were more intelligent than non-wearers.

The two researchers made videotapes of performers both wearing and not wearing glasses. They presented them to the audience either as a static 15-second picture or as a video clip in which performers talked on a basic, non-intellectual topic – how they were going to spend their holidays – for 5 minutes. Judges rated the spectacle wearers 12 points higher in IQ than the non-wearers, the question relating to IQ being buried in a mass of other camouflaging ones. This IQ 'difference', however, only existed when the viewers looked at wearers in static photographs. Once the 'spectacles' opened their mouths the magical effects disappeared. No real IQ differences existed.

And then there's the doctor who looks at the intestines of bad guys.

⚛ He's got guts

Francesco Aragona of the University of Messina in Sicily spent 30 years dissecting the victims of shoot-outs between mafiosos. In 1989, he said he could help police tell the difference between culprits and innocent bystanders just by looking at internal organs. In the gangsters, these were ulcerated, enlarged and otherwise showing signs of extreme stress. Aragona's Roman predecessors predicted the future by looking at the organs of slaughtered animals. Were they on to something?

Of course, in the end, all medical science just boils down to a single, ultimate quest; the search for the perfect hangover cure. But that couldn't be left to fancy-dan professors. In 1999 New Scientist *got experimental itself and sent its own team of volunteers into action.*

✳ Desperate remedies

All but the saintliest of us have had them. Hangovers approaching near-death experiences. Raging thirst, thumping headache, wobbly limbs and nausea. Worse, there's no magic bullet to make us instantly well. Thousands of years after the first 'morning after the night before', we can send people to the Moon and create computers of mind-numbing power, yet we are still far away from a science-based, experimentally verified hangover cure. Why? The simple answer is that in the eyes of most governments, doctors and industries, a hangover cure would trigger a catastrophic upsurge in alcohol abuse, tempting mild drinkers to overindulge. Hangovers are, after all, nature's way of saying 'don't do this to yourself'.

So don't expect to find the magic bullet nestling anywhere in the scientific literature. The research hasn't been done nor is it ever likely to be done. But there is plenty of research on what alcohol does to the body. Armed with this knowledge, many alcohol researchers have speculated about the best hangover remedies.

In the interests of easing global pain, *New Scientist* decided to test some of the top tips to emerge from this research. A panel of a dozen or so intrepid volunteers agreed to overindulge on four successive weekends, trying a different 'cure' each time. The morning after, they recorded how lousy they felt by scoring a range of symptoms. Because dehydration is the most well-documented consequence of drinking, whichever concoction they tried, the volunteers drank a pint or so of water before going to bed.

And the truth of the matter? Overall, nothing worked 100 per cent of the time, highlighting just how complex hangovers are. 'Hangovers are multifactorial,' said Thomas Gilg, an alcohol researcher at the University of Munich's department of forensic chemistry. 'If you smoke, or haven't had enough sleep, that can affect the hangover, too.'

So, we will have to keep waiting for that elusive magic bullet. Most sore heads will have to be soothed with the traditional remedies our researchers tried – plenty of water before bed, something sugary, plus cysteine-rich foods such as eggs, which help mop up any damaging chemicals. To take the edge off headaches, pop an aspirin or ibuprofen before bed. But avoid paracetamol, known in the US as acetominophen, which according to the US Food and Drug Administration amplifies alcohol's damaging effect on the liver.

The more adventurous could probably do worse than consider a combination of all the cures our volunteers tested – water, sports drinks, cysteine and, for real emergencies, a vodka-based pick-me-up. 'It's a bit of folk medicine, but it would probably do you no harm,' said Waltenbaugh. But the best cure, of course, is not to drink at all. Cheers!

5 Blunders – big and small

'To lose one parent, Mr Worthing, may be regarded as a misfortune; to lose both looks like carelessness.' So observed Lady Bracknell in Oscar Wilde's *The Importance of Being Earnest*. It's certain that neither the fictional Mr Worthing nor the equally fictional Lady Bracknell had ever met Stephen Myers, who worked with the Large Electron Positron Collider in Geneva. But presumably he'd empathise with the plight of Mr Worthing. After all it wasn't just one, but two carelessly lost empty bottles of beer that caused his collider to, er, fail to collide after its very expensive upgrade.

Oh, it's easy to mock, of course. Which is probably why we are devoting a whole chapter to the art. But if you let scientists do exactly what they want, sometimes things are going to go awry. And it's not just discarded beer bottles left inside expensive scientific equipment that cause problems. Not knowing the difference between imperial and metric units can have serious consequences for your spacecraft and a liking for the taste of lead seemingly does not help you quit smoking.

While we're on the subject of mockery, it is instructive to recall that the art has a long history in the field of science. Back in the 1890s, before the nature of electricity was truly understood, an aspiring student was sitting an oral exam at Oxford University. He was asked if he knew what electricity was. 'I'm sure I've learnt what it is,' stuttered the nervous candidate, 'but now I've forgotten.' 'How very unfortunate,' replied the unimpressed examiner. 'Only two people have known what electricity is, the Author of the Universe and yourself. And now one of them has forgotten.' Clearly the examiner was being a little unfair on his minion, which is perhaps what we'll be accused of doing with this trawl through the pitfalls of any number of keen but ultimately unfortunate scientific endeavours.

However, a charge of 'being a little unfair' cannot be levelled at us when we consider the case of Pierre Pumpille, who provides one of the stories that fail to make it into this chapter. In a feature on the excesses that men will go to in order to prove their machismo, *New Scientist* told the story of the headstrong Frenchman who had shunted a stationary car a distance of 2 feet by headbutting it. 'Women thought I was a god,' he explained from his hospital bed. We disagree, and despite using his head in a way that scientists usually do not in order to prove his point, he fails to make the grade. For while stupid, his research cannot be declared a 'blunder'. He actually meant to headbutt the car.

We have also avoided mention of Pumpille's fellow countryman Michael Fournier in this chapter. Monsieur Fournier has for many years been attempting to beat the record for the highest freefall parachute jump. His latest mishap occurred in 2008 when the balloon that was set to carry him to the world altitude record departed without him. That might, in most people's books, constitute a strong case for inclusion in this chapter, but we are not so presumptuous. We are going to wait until the next edition before we pass judgement, by which time we are certain Michael will have achieved his lofty aim.

Scientists sometimes seem so clever that it's perhaps more than just schadenfreude that makes us want to see them taken down a peg or two. Fortunately, there are plenty of examples of their failures around to make us feel a little better about ourselves. Interestingly, rocket science, that clichéd branch of all that is über-brainy, seems to suffer more than its fair share of calamities.

☀ The trouble with rockets

Rocket science has a deserved reputation for being tough, so it should be no surprise that things can get a little bumpy when it comes to designing and testing a launcher. Just ask Elon Musk, the millionaire founder of PayPal and rocket company SpaceX.

His first rocket, *Falcon 1*, was scheduled to lift off from the Pacific atoll Kwajalein in November 2005, at which point the problems began. First, unplanned engine tests used up more liquid oxygen (LOX) fuel than expected, then the LOX generator broke down. A shipment of fuel was ordered from Hawaii, but the tanker sprang a leak and arrived only one-fifth full. There was just enough to launch, but a valve was left open during the final preparations, allowing more fuel to waste away, so the launch had to be cancelled.

A fresh shipment of LOX arrived a month later but disaster struck again. High winds hit the atoll, and to be on the safe side, engineers decided to drain the fuel from the rocket. A faulty pressure valve caused a vacuum to form inside the main fuel tank, sucking in its soft sides like a crushed beer can.

After three months of repairs, by March 2006 *Falcon 1* was ready to go. Seconds after launch, however, the main engine sprang a fuel leak, leaving a trail of flame in the rocket's wake as it spiralled off course and crashed within sight of the disappointed engineers. An investigation pinpointed the cause: a corroded fuel-line nut was to blame.

A year later, with the rocket rebuilt from scratch, *Falcon 1* finally took off without a hitch. It managed 5 minutes of smooth flight. But a bump as the first and second stages separated confused a control system, causing it to enter an uncontrolled roll, which triggered a premature shutdown of the second-stage boosters. *Falcon 1* did reach space, but not with the velocity needed to secure orbit.

Its next outing was scheduled for January 2008, when it was to carry the cremated remains of 125 people, including actor James Doohan – *Star Trek*'s Scotty.

Unfortunately, when Scotty finally blasted off on 2 August 2008, Falcon 1 failed again, bringing his ashes down along with those of 207 other cremated people. Undeterred (or perhaps uninterred), the rocketeers intend to keep on trying. At least they are not alone ...

Schoolkid blunder brought down Mars probe

In 1999, NASA lost its $125 million Mars Climate Orbiter spacecraft as a result of a mistake that would shame a first-year physics student – failing to convert imperial units to metric. The problem arose from a culture clash between spacecraft engineers and navigation specialists, said Mary Hardin, a spokeswoman for NASA's Jet Propulsion Laboratory in Pasadena. 'Propulsion people talk in pound-seconds of thrust and navigators talk in newton-seconds,' she said.

The spurious data came from the craft's attitude-control system, a design which had worked fine on the Mars Global Surveyor. But there was one crucial difference in the system on the orbiter. 'There was a different propulsion supplier for the Mars Climate Orbiter, and its data package was in imperial units,' said Noel Hinners, vice-president for flight systems at Lockheed Martin Astronautics in Denver, Colorado. No one adapted this data-processing software for the second probe, so JPL's navigation software thought the numbers it received were newton-seconds rather than pound-seconds. The attitude thrusters only made small corrections, but the error was enough to leave the probe 100 kilometres too close to Mars when it tried to enter orbit.

It's not just space scientists, of course, who screw up big time. The air force, the navy, they're all at it …

✸ Back to the future

In 2007, 12 F-22 Raptors, the US Air Force's new stealth fighters, left Hickam Air Force Base in Hawaii, bound for Okinawa, Japan, on the high-tech planes' first overseas outing. Things went smoothly until they reached the 180th meridian – otherwise known as the International Date Line.

Some of the pilots suddenly found themselves without any navigation aids. With nothing to tell them their compass heading or even whether they were level or not, it was as if the pilots had been instantaneously transported from the cockpit of the world's most advanced aircraft into that of one dating from the First World War. Fortunately, the skies were clear, so the squadron did an about-face and was able to follow its in-flight refuelling tankers back to Hickam.

The error was diagnosed as a problem with a 'partial line of code' that had pitched the planes' computers into an infinite loop of trying and failing to calculate their position while dealing with an unexpected date. A fix was issued, and 3 weeks later the planes made their trip to Japan without a hitch.

'Reliance on electronics has changed the flight-test process,' said Donald Shepperd, once head of the US Air National Guard. 'It used to be tails falling off, now it's typos that ground a fighter.'

❊ The torpedo is not for turning

In 1993, a torpedo owned by the Institute of Antarctic and Southern Ocean Studies at the University of Tasmania disappeared after heading in the wrong direction under the ice off Antarctica.

The torpedo was a research tool taken to Antarctic waters on board the *Aurora Australis*, the research vessel of the Australian Antarctic Division. The ship returned to Hobart without the torpedo and the A$35,000 worth of computer equipment it carried. On its maiden voyage, the torpedo failed to return as it was programmed to.

The obsolete weapon was bought from the Royal Australian Navy for A$150 and fitted with sonar equipment. The aim was to bounce signals off the bottom of the ice pack to measure its depth. It was programmed to do a U-turn under the ice before being retrieved. In open water, the device performed perfectly, but under the ice it failed the test. 'There was one bright spot,' said Garth Paltridge, director of the institute. 'Part of the exercise was to test the logistics of operating in such a hostile environment. In that sense it worked perfectly.'

After being lowered into the water and before its motor started, the waves turned the torpedo about 90 degrees from its course. Its guidance system never brought it back on track. Also, it was so cold, about -20 °C, that the ship's tracking system stopped working.

Yet, while it's clear other organisations also make mistakes, NASA can often be seen lurking in the background.

Memory scramble

They say ballooning is the least stressful way to fly. Indeed, a balloon seemed the perfect platform from which the $10 million BLAST telescope, funded by NASA, the Canadian Space Agency and the UK's Particle Physics and Astronomy Research Council, could take far-infrared snaps of star formations.

For 12 days running up to 2 January 2007, the telescope collected valuable data as it floated 40 kilometres above Antarctica. As it descended, the gondola released the huge balloon and deployed its landing chutes as planned. But the electronics that should have released the parachutes on touchdown failed. Antarctic winds inflated them like giant spinnakers, turning the cargo into a wind-powered sled.

'It was moving as fast as I could run,' recalled project leader Mark Devlin, who was following the fate of the 1800-kilogram telescope and its support computers from his home in the US. 'It was absolutely sickening.' The support plane could only watch as the gondola bounced across the ice, strewing pieces of equipment as it went. It finally came to rest 24 hours later, when it wedged in a crevasse 200 kilometres from its landing site.

Devlin was emailed a picture of the scene, which he scrutinised for any sign of the hard drives which bore the only copy of the mission's data. 'NASA paints everything white,' he said, so his search was initially in vain. Fortunately, a pilot tracking the furrow gouged by the gondola spotted the package. The damaged drives eventually yielded their irreplaceable data, but the telescope was a write-off. Devlin then planned to fund-raise for a similar mission, with tougher electronics and one other change: 'I'm thinking fluorescent orange,' he mused.

✳ Robot's return

In 1993, NASA robotic experts suffered a setback when an experimental robot broke down during a highly publicised demonstration in an Antarctic volcano. Following that, experts planned to try again, with an improved version of the robot and in a different volcano. If all went according to plan, Dante II was to explore the crater of Mount Spurr, a volcano in Alaska. The eight-legged robot was studded with video cameras, a laser range finder and sensors that would guide its motion.

Dante II also had an improved communications tether linking it to its human overlords. The tether on the original Dante broke during its test at Mount Erebus in Antarctica in January 1993. The project was intended to test and demonstrate technologies that could be used in remotely controlled space probes. Dante II was to descend about 600 feet into the volcano's crater and record temperatures and the make-up of gases there.

So, what happened on its second expedition?

✳ Dante rescued from crater hell

NASA'S accident-prone robot Dante II was hoisted out of an Alaskan volcano in 1994, more than a week after it lost its footing and tumbled into the volcano's mouth.

Dante II's Alaskan mission made a promising start. The robot trekked down into the crater of Mount Spurr and sampled gas from an open vent. But as Dante was climbing back out of the crater, it lost its footing in the slippery volcanic mud. It fell on its side and, like a giant beetle, was unable to move.

In the first rescue attempt, a helicopter tried to lift Dante out of the crater by the fibre-optic cable linking the robot to its controllers. The cable snapped, and Dante fell further into the volcano. Further helicopter rescue missions were hampered by fog. Eventually, two workers climbed down into the crater and attached a harness to the robot's body so that Dante could be winched out by helicopter.

Things didn't get much better, did they? Dante was behaving like the embarrassing drunk uncle at the family party. And scientists know full well that alcohol and serious research sometimes don't mix.

Particle beams hit the bottle

Physicists starting up Europe's most powerful particle accelerator in 1996 hoped to detect fundamental particles not seen since the beginning of the universe. But when the Large Electron Positron (LEP) collider at the CERN particle physics laboratory in Geneva resumed operation after a £210 million upgrade, nothing happened. The fault, it emerged, was due to two empty bottles of lager.

LEP sits in a 27-kilometre circular tunnel. In the 1980s, physicists used the accelerator to prove the existence of the W and Z particles, which carry the weak nuclear force. With the energy of LEP's particle beams nearly doubled to 162 giga-electronvolts, the accelerator had a chance of uncovering the Higgs boson, which was thought to endow other particles with mass.

When LEP2, as the upgraded machine was known, was switched on, beams of electrons and positrons were supposed to travel in opposite directions round the giant circuit, but neither beam made it. CERN's investigators homed in on the section of the ring causing the trouble. When they dismantled it, they found that the high-vacuum tube through which the beams travel had been blocked by two beer bottles.

No damage had been done to LEP2. But the incident had a scientific cost. 'We lost five valuable days of physics,' said Stephen Myers, who was in charge of LEP2.

✳ A matter of perspective

It's reassuring to know that an engineering screw-up doesn't always get you into trouble. It can sometimes even dig you out of it, as astronomers found out at the Canada-France-Hawaii Telescope (CFHT) on Mauna Kea, Hawaii.

In early 2003, researchers began observing the skies using the telescope, equipped with a new digital camera – the biggest in the world at the time. The CFHT was fitted with four precision lenses so the camera could capture crisp images of vast areas of sky.

The results were disappointing. The images were sharp in the centre, but far more blurred than expected at the sides.

A laborious investigation followed, with engineers dismantling the optics and reassembling them daily, but finding no answer. Then one day, an engineer mistakenly replaced one of the four lenses back-to-front. The images improved spectacularly.

'The next observations were just "Wow!",' said Christian Veillet, the observatory's director. 'The image quality was just what it should have been.'

No one understood why the back-to-front lens worked so well, or why it didn't work when it was oriented as planned. 'That has been frustrating, but it would be a waste of resources to investigate, so we decided to just forget about it,' said Veillet. 'Now the science that is coming out is exquisite.'

To be fair, it's easy to criticise. How many great ideas have you had today? They too may have ended up in a crater like Dante, or an Antarctic crevasse like BLAST. So thank your lucky stars that no one has taken yours seriously. Somebody obviously thought it reasonable to take these researchers from the University of Oklahoma seriously, and look what happened. Whoever thought it was a good idea was just plain, unacceptably stupid.

⚛ Elephants on acid

What happens if you give an elephant LSD? Researchers solved this mystery in August 1962, when Warren Thomas, director of Lincoln Park Zoo in Oklahoma City, fired a cartridge-syringe containing 297 milligrams of LSD into the rump of Tusko the elephant. With Thomas were two colleagues from the University of Oklahoma School of Medicine, Louis Jolyon West and Chester M. Pierce.

The dose was about 3000 times what a human would typically take. Thomas, West and Pierce figured that if they were going to give an elephant LSD they'd better not give it too little. They later explained that the experiment was designed to find out if LSD would induce musth in an elephant – musth being a kind of temporary madness male elephants sometimes experience during which they become highly aggressive and secrete a sticky fluid from their temporal glands. One may also suspect a small element of ghoulish curiosity was involved.

Whatever the reason for the experiment, it almost immediately went awry. Tusko reacted as if he had been shot by a gun. He trumpeted around his pen for a few minutes and then keeled over. Horrified, the researchers tried to revive him with a variety of antipsychotics, but about an hour later he was dead. In an article published four months after the event in *Science*, the three scientists sheepishly concluded: 'It appears that the elephant is highly sensitive to the effects of LSD.'

The experiment instantly made headlines. Faced with a public relations disaster, the scientists protested their innocence. They had not anticipated the elephant would die, they insisted. In their experience, LSD was a powerful hallucinogen but rarely fatal. West and Pierce helpfully noted that they themselves had previously taken the drug.

Thomas tried to find a silver lining. They had learned that LSD can be lethal to elephants. So perhaps, he mused, the drug could be used to destroy herds in countries where they are a problem. For some reason, his suggestion has never found any takers.

We can only hope that the Oklahoma researchers copped all the opprobrium they were due, but for some people – despite their own stupidity – you have to feel a little sympathy. The following story is, we freely admit, not a tale of a researcher being given free rein. But it does illustrate a blunder in the use of technology that we have all surely come across ourselves. How many times have you hit 'Reply All' to an email and then thought 'Aaaaaaaaargh'?

Naming and shaming

In 1993, the NatWest Bank admitted that it kept personal information about its customers – such as political affiliation – on computer. But *Computer Weekly* revealed that a financial institution, sadly unnamed, went one better and moved into the realm of personal abuse.

The institution decided to mailshot 2000 of its richest customers, inviting them to buy extra services. One of its IT workers wrote a computer program to search through its databases and select the customers automatically. He tested the program with an imaginary customer called Rich Bastard.

Unfortunately, an error resulted in all 2000 letters being addressed 'Dear Rich Bastard'. The luckless programmer was subsequently sacked.

✳ Target practice

A good story deserves a good airing. Back in the 1990s, two members of the Lothian and Borders traffic police were having a pleasant time out on the Scottish moors, trapping speeding motorists with a radar gun. Suddenly their equipment went crazy, registering a speed of more than 300 miles per hour. It then locked up completely.

Seconds later, the startled cops understood why, as a low-flying Harrier jet screamed over their heads. Upset that their radar gun had been broken, the policemen put in a complaint to the Royal Air Force – only to discover that the damage could easily have been so much worse.

The RAF informed them that the Harrier's target-seeker had locked on to what it had interpreted as enemy radar. This immediately triggered an automatic air-to-surface missile attack. Fortunately for the two policemen, the Harrier was operating unarmed.

The policemen didn't blunder, neither did the fighter pilot. But a software developer must now be wiping a brow. They might want to compare notes with the designer of the Swedish navy's early-warning system.

✴ Furry submarines embarrass Swedish navy

When is a sub not a sub? The answer, as an embarrassed Swedish navy had to admit in 1996, is when it's a mink or an otter. What the navy thought was the sinister sound of Soviet propellers was, in fact, the furious paddling of little legs.

A scientific commission set up by the government and chaired by the former director of the Swedish Engineering Science Academy, Hans Forsberg, concluded that most of the invading submarines reported by the navy were mythical. Of more than 6000 reports of 'alien underwater activity' between 1981 and 1994, the commission found firm evidence for only six incidents.

In every other case the evidence, often based on sightings by the public, was unreliable. The navy claimed that one underwater noise, similar to the sound of frying eggs, was caused by bubbles from submarines. It was more likely to have been caused by the natural movement of water, said the commission.

On 40 occasions between 1992 and 1994, a defence network of microphones attached to buoys detected the sound of bubbles caused by a rotational movement in the water. The navy estimated the speed at up to 200 revolutions per minute, and assumed it must be submarine propellers.

But the navy was wrong, said the commission. According to its secretary, Ingvar Akesson, tests with swimming mink or otters showed that they could produce the same readings as propellers. 'It is very puzzling, but they do,' he said. The navy accepted that in some cases it might have confused the two sounds.

And it's still the defence of the realm to which we turn for an even more laughable instance of inventor stupidity.

✳ 'Secret' terrorist trap gets worldwide publicity

In 1991, two companies jointly developed a security system for airports designed to trap terrorists and drug smugglers. To be effective the system, which detected traces of bomb-making chemicals on passenger boarding cards, relied on complete secrecy – but to protect their commercial interests the developers made the mistake of filing a European patent application, thereby ensuring that full details were available in libraries all around the world for everyone, including terrorists and drug smugglers, to read.

Paul Wilkinson, director of the Research Institute for the Study of Conflict and Terrorism at St Andrew's University, confirmed that publishing the idea in a patent application gave the game away. 'The more sophisticated organisations read all the scientific literature,' he said. 'They study the technology and work out ways of evading it. Now it will only pick up mad bombers with home-made devices and amateur newcomers,' he feared.

All of which leads to a final practical example of direct experimentation. There can be no research more trustworthy than that which has been carried out by the subjects themselves. Although not a scientist, this lay researcher will be thanked by anybody attempting to quit smoking for proving to them what they might have suspected in the first place.

✳ Better lead than dead

If you are thinking of giving up smoking, you are recommended not to chew electrical cables to ease your withdrawal pangs. This, according to the *Australian Medical Journal*, is what one Antipodean building site worker did. He got through almost a metre of cable a day for ten years, and found it extremely soothing, with a 'sweet and pleasant taste'.

In 1996, the man was admitted to hospital with stomach pains. Unfortunately, the cables he was gnawing contained lead to make them more flexible. And doctors found that the lead levels in the man's blood were three times the safety limit.

Thanks to prompt treatment, his blood level returned to normal. But he failed to quit smoking.

6 Love, sex and all that stuff

Ah yes … sex. It's probably fair to say that scientists are rarely trendsetters. So if sex wasn't invented until the 1960s, science didn't truly embrace this groundswell of fashion until a couple of decades later, as our pick of the bunch below seems to indicate. But once scientists got their teeth into it, just as with any discipline, there was going to be no lying back and thinking of Einstein.

It's possible that scientists were desperate to prove that they were as normal (or at least as open-minded) as the next person, a process that perhaps reached its apotheosis when, in the mid-1990s, an American science radio show thought it prudent to release the annual beefcake calendar *Studmuffins of Science*. Science has been apologising ever since.

But in the noughties we are all a little more grown up than we were. Or at least we hope we are. Nonetheless, some of the items we discovered when researching this book raised eyebrows (if very little else). All of which meant that deciding what to put into this chapter and what to leave out left the editor in a similar dilemma to the one he faced when, on selecting the first seven letters in a game of Scrabble he drew out, at incredible odds and in straight order: S-C-R-O-T-U-M. The word would almost certainly win the match on the first play of the game but could you – dare you – impose the word on your 80-year-old grandmother?

Such mental torment loomed over this chapter. *New Scientist* has certainly covered some esoteric sex-based discoveries over the years, but should we really inflict the Hungarian musical condom on an unsuspecting readership? Or the science of *Baywatch*? Find out which one made it and which was consigned to the dustbin of tack below ...

Oh, and don't read this if you're under the age of consent. And if you have no idea what that is, ask your mum and dad, not a scientist ...

We admit we are a bit boring, but we think sex should be enjoyable. So we thought it was good news for women around the world when New Scientist *announced an unexpected discovery.*

✳ Push my button

In the Woody Allen comedy *Sleeper*, a machine called an orgasmatron delivers an orgasm at the push of a button – without the hassle of sex. In 2001, life imitated art when scientists in the US patented an implant that achieved the same effect for women whose lives were blighted by an inability to achieve orgasms naturally.

Orgasmic dysfunction is not uncommon among women, said Julia Cole, a psychosexual therapist and consultant with Relate, the relationship counselling service. And a number of issues can cause it, said Jim Pfaus, who studies the neurobiology of sexual behaviour at Concordia University in Montreal. 'Some women confuse what's called sympathetic arousal, like increased heart rate, clammy hands, nerves and so on, with fear,' he explained. 'That makes them want to get out of the situation.' Psychotherapy is a common treatment for the condition, although if anxiety is a factor, patients may also be prescribed Valium. 'But Valium can actually delay orgasm,' said Pfaus.

Stuart Meloy, a surgeon at Piedmont Anesthesia and Pain Consultants in Winston-Salem, North Carolina, got the idea for an orgasm-producing device while performing a routine pain-relief operation on a woman's spine. 'We implant electrodes into the spine and use electrical pulses to modify the pain signals passing along the nerves,' he said. The patient remains conscious during the operation to help the surgeon find the best position for the electrodes. Meloy's breakthrough came one day when he failed to hit the right spot. 'I was placing the electrodes and suddenly the woman started exclaiming emphatically,' he said. 'I asked her what was up and she said, "You're going to have to teach my husband to do that".'

Meloy said the stimulating wires could connect to a signal generator smaller than a packet of cigarettes implanted under the skin of one of the patient's buttocks. 'Then you'd have a hand-held remote control to trigger it,' he said. 'But it's as invasive as a pacemaker, so this is only for extreme cases.'

Meloy believed the device could help couples overcome problems caused by orgasmic dysfunction. 'If you've got a couple who've been together for a while and it's just not happening any more, maybe they'll get through it a bit easier with this,' he said.

He was quick to add that the device would be programmed to limit its use. 'But whether it's once a day, four times a week – who am I to say?'

But would women subject themselves to such invasive surgery? 'If young women of 15 or so are having painful operations to enlarge their breasts when they don't have to, are you kidding? Of course it'll be used,' said Pfaus.

Surprisingly, Jim Pfaus was wrong …

✳ Wanted: people to test orgasmatron

Clinical trials of the 'orgasmatron' began in the US in 2003, with the approval of the Food and Drug Administration.

The device was the focus of massive media attention in 2001, after *New Scientist* broke the news of its existence and used the term orgasmatron to describe it. But despite all the coverage, few people volunteered for the trial. 'I thought people would be beating my door down to become part of the trial,' said Meloy. 'But so far I am struggling to find people.'

That did not surprise some experts, who thought an implant was too radical a treatment for sexual problems. 'Why would you do it invasively if you can do it with a vibrator?' asked Marca Sipski of the University of Miami, who studied sexual function in women with spinal cord injuries.

By 2003 only one woman had completed the first stage of the trial, and just one other had been signed up. Meloy hoped to find eight more to complete the first stage of the study, in which wires connected to a battery pack were inserted through the skin and into the woman's spinal cord. The procedure is no riskier than an epidural, Meloy said. But epidurals can cause complications such as backache in up to a fifth of women. In the second stage, a self-contained device resembling a pacemaker would be implanted beneath the skin, and switched on and off with a remote control. Meloy expected a full implant to cost around $13,000.

He implanted wires in a woman who responded to his call for volunteers in the local media. 'When the device was switched on, the patient reported being almost instantly aroused. She described it as "really excellent foreplay",' said Meloy. The woman, who had not had an orgasm for four years, wore the device for nine days and had sex with her husband on seven occasions. Meloy said she had an orgasm every time. 'She even told me she had the first multiple orgasm of her life using the device,' he said.

But Sipski thought that as long as the required nerves in the body were intact, using a vibrator should work just as well. 'My research shows that orgasm is a purely reflex response. Even the sensation associated with orgasm does not require the brain. Women with complete injuries to the spine can still experience orgasm.'

Hopefully, the following discoveries will prove to be more popular or at least more easily accessed.

❊ Ecstasy over G spot therapy

It has evaded lovers for centuries, but in 2008 the elusive and semi-mythical G spot was captured on ultrasound for the first time.

Emmanuele Jannini at the University of L'Aquila in Italy discovered clear anatomical differences between women who claimed to have vaginal orgasms – triggered by stimulation of the front vaginal wall without any simultaneous stimulation of the clitoris – and those that didn't. Apparently, the key is that women who orgasm during penetrative sex have a thicker area of tissue in the region between the vagina and urethra, meaning a simple scan could separate out the lucky 'haves' from the 'have-nots'.

Even better, Jannini found evidence that women who have this thicker tissue can be 'taught' to have vaginal orgasms. Ultrasound scans on 30 women uncovered G spots in just eight of them and when these women were asked if they had vaginal orgasms during sex, only five of them said yes. However, when the remaining three were shown their G spots on the scan and given advice on how to stimulate it, two of them subsequently 'discovered' the joy of vaginal orgasms. 'This demonstrated, although in a small sample, the use of [vaginal ultrasound] in teaching the vaginal orgasm,' Jannini said.

Sadly, none of the have-nots had vaginal orgasms either before or after the scans, so they would just have to make do with the old-fashioned clitoral kind.

Jannini went on to investigate whether hirsute women are more likely to have G spots since they have higher levels of testosterone and both the clitoris and the G spot are thought to respond to the hormone. The burning question is whether women with a small G spot can 'grow' it with practice. Jannini was optimistic. 'I fully agree that the use makes the organ. I do expect an increase with frequent use.' So perhaps the only way to make the most of your G spot, if you have one, is to get practising.

Ultrasound listens in to love in the lab

Sexual penetration is best from behind or from the side, claimed a unique study of the act of human copulation. The study, conducted by scientists at the University Hospital in Copenhagen, Denmark, in 1991, showed that those sexual positions which achieve maximum penetration by the male also result in the most climactic orgasm for the female and the best chance of fertilisation.

The findings suggested that several traditional western sexual positions were unfulfilling to women and the least likely to result in pregnancy. 'It's all basically a question of the shape of the erect penis in relation to the woman's vagina and other sexual organs,' explained Lasse Hessel, the leader of the research team. The deeper the penetration, the more profound the resultant orgasm and hence the greater the chance of ejaculated sperm fertilising an egg.

Hessel and his team based their conclusions on extensive observations of Danish couples making love in the laboratory, using ultrasound machines normally used to monitor the development of the foetus in pregnancy. The scans revealed the exact position and movement of the penis during intercourse.

Hessel and his colleagues scored positions on several objective criteria, such as depth of penetration and how long it took each couple to reach orgasm, and on subjective criteria such as each partner's rating of the intensity of their orgasms.

The positions ranged from the most popular 'missionary' position to the less popular sideways or rear-entry positions. 'The rear and sideways positions scored best in virtually all cases,' Hessel concluded. These positions, which enabled the curve of the erect penis to match that of the female anatomy, also maximised the chance of the woman experiencing multiple orgasms.

But not everybody was happy.

✳ Fertile orgasms

Sir, I was very interested to read the article entitled 'Ultrasound listens to love in the lab' and would love to know more details of these experiments. The results appeared to fly in the face of research such as the Hite Report on women's experience of sex.

Over the past decade or so it has become recognised that women have a right to express their requirements as far as lovemaking is concerned, and that this does not usually entail just being banged into from behind with maximum penetration. It has been well established that orgasms of mammoth proportions can be produced from clitoral stimulation, without any need for penetration by a penis at all. Was there any clitoral stimulation during the experiments you quote? If not, how long did it take before these women reached orgasm, as I should think it would be longer than most men could manage.

If a women has already reached orgasm once then maybe I can believe what the report says, but in that case there has been a crucial omission in your account. If there's no such omission then please, tell us how it's done.

Elizabeth Boothman
(18 January 1992)

If you agree, and feel strongly that the Copenhagen researchers are misguided, there are – thankfully – still more ways to achieve orgasm. Trouble is, this one involves being watched too ...

⚛ Hot news

In 1980, psychologists Lisa Berry and Paul Abramson at the University of California at Los Angeles hoped they had discovered a new aid for people with sexual problems. Using a thermography machine of the type used to look for the hot spots (or changes in infrared radiation) caused by tumours, they were able to help cure patients who claimed they were unable to achieve orgasm.

Patients were encouraged to masturbate in front of a TV camera, and the flow of blood and energy to the erogenous zones showed up as changes in colour on a screen in front of them. The fact that they could see that they were turning themselves on helped them to obtain confidence in their ability to stimulate themselves successfully, and orgasm was often achieved during the first or second session – sometimes the patients claimed for the first time ever in their lives.

Still, as we are increasingly learning, it takes all sorts …

⚛ Safe sex

A 1980 issue of the *British Medical Journal* carried an intriguing report from two surgeons, one from University College Hospital London and one from St Thomas', also in London. They wrote on 'Penile injuries from vacuum cleaners' – a title guaranteed to stop any contents-skimmer in his or her tracks.

The two doctors were obviously men of the world; they introduced their account of woeful damage suffered by the male organ when inserted into a vacuum cleaner nozzle with the prefatory remark that the practice was undertaken 'probably in search of sexual satisfaction'.

Neil Citron and Peter Wade reported four cases of damage after attempted copulation with a machine. But far more interesting than the details of the damage is the fact that none of the men was exactly honest about his proclivities …

⚛ Spanking 'brings couples together'

Spanking is stressful at first, but it could bring consenting couples closer together. That's the implication of two studies conducted in 2009 of hormonal changes associated with sadomasochistic (S&M) activities including spanking, bondage and flogging.

Brad Sagarin at Northern Illinois University in DeKalb and colleagues measured levels of the stress hormone cortisol in 13 men and women at an S&M party in Arizona, before, during and after participating in activities. During S&M scenes, cortisol rose significantly in those receiving stimulation, but dropped back to normal within 40 minutes if the scene went well. There was no change in those inflicting the activity.

At an S&M event in Colorado, testosterone was measured in 45 men and women. It increased significantly in receiving women only. Donatella Marazziti of the University of Pisa, Italy, said the boost may help women cope with the aggressive nature of S&M activities, or that it could be another sign of stress. In both studies, couples who said the party went well also reported increases in relationship closeness.

It's important to note that levels of both hormones dropped back down in couples who enjoyed the experience, Marazziti said. 'When sexual intercourse is consensual it is not stressful – even if it is extreme sex.' Richard Wiseman, a psychologist at the University of Hertfordshire in Hatfield, UK, added that almost any shared activity is likely to promote interpersonal closeness. 'It doesn't have to be tying up your partner or placing clamps on their nipples, it could be something as simple as cooking a meal together or even doing the housework as a duo,' he said.

Nick Neave, a psychologist at the University of Northumbria, UK, said the results were interesting, but future studies should control for whether participants experienced orgasm, which is associated with reduced stress and an increase in hormones associated with partner-bonding and affection.

But we are a little old-fashioned and, given the choice between a flogging and a wooing, we'd opt for the latter. In 2006, New Scientist *took a look at the best ways to woo.*

✳ Six ways to woo your lover

Let your body do the talking

We all hunt for the perfect chat-up line, but in reality, our bodies give away a great deal before we open our mouths. It is estimated that when you meet a stranger, their impression of you is based 55 per cent on your appearance and body language, 38 per cent on your style of speaking and a mere 7 per cent on what you actually say. So what can we learn from the experts? There are a number of actions that signal 'I like you' to another person. Adopting an open posture (no folded arms) and mirroring another's posture help create a feeling of affinity. Most people are not conscious of being mirrored, but evaluate those who do it more favourably. And it is worth adopting stances that enhance your masculinity or femininity, such as placing hands in pockets with elbows out to enlarge the chest. You could also indulge in a 'gestural dance', synchronising your gestures and body movements with those of the object of your desire, such as taking a sip of your drinks at the same time.

Experience fear together

A dramatic setting can kick-start your love life. Meeting a stranger when physiologically aroused increases the chance of having romantic feelings towards them … It's all because of a strong connection between anxiety, arousal and attraction. In the 'shaky bridge study' carried out by psychologists Arthur Aron and Don Dutton in the 1970s, men who met a woman on a high, rickety bridge found the encounter sexier and more romantic than those who met her on a low, stable one. A visit to the funfair works wonders too. Photos of members of the opposite sex were more attractive to people who had just got off a roller-coaster compared with those who were waiting to get on. And couples were more loved-up after watching a suspense-filled thriller than a calmer film.

Why? No one is sure, but the adrenaline rush from the danger might be misattributed to the thrill of attraction. But beware: while someone attractive becomes more so in a tense setting, the unattractive appear even less appealing.

Share a joke
An experience that makes you laugh creates feelings of closeness between strangers. A classic example comes from experiments carried out by psychologists Arthur Aron and Barbara Fraley, in which strangers cooperated on playful activities such as learning dance steps, but with one partner wearing a blindfold and the other holding a drinking straw in their mouth to distort speech. Sounds stupid, but love and laughter really did go together. We suggest that the blindfold/ drinking straw approach is best confined to the laboratory.

Get the soundtrack right
Psychologists at North Adams State College in Massachusetts have proved what Shakespeare suggested – that music is the food of love. Well, rock music, at least. Women evaluating photos of men rated them more attractive while listening to soft-rock music, compared with avant-garde jazz or no music at all.

Gaze into their eyes
Any flirt knows that making eye contact is an emotionally loaded act. Now psychologists have shown just how powerful it can be. When pairs of strangers were asked to gaze into each other's eyes, it was perhaps not surprising that their feelings of closeness and attraction rocketed compared with, say, gazing at each other's hands. More surprising was that a couple in one such experiment ended up getting married. Neuroscientists have shed some light on what's going on: meeting another person's gaze lights up brain regions associated with rewards. The bottom line is that eye contact can work wonders, but make sure you get your technique right:

if your gaze isn't reciprocated, you risk coming across as a stalker.

Use love potions?
Can you short-cut all the hard work of relationship-building by artificial means? People have been trying to crack this one for thousands of years. A nasal spray containing the hormone oxytocin can make people trust you – an important part of any relationship – though there's no evidence yet to suggest it can make someone fall in love. And while we wouldn't suggest you try this at home, studies on prairie voles show that injecting the hormone vasopressin into the brain makes males bond strongly to females. Illegal drugs such as cocaine or amphetamines can simulate the euphoria of falling in love by raising levels of the neurotransmitter dopamine, but dopamine levels can also be increased legally by exercising. Another neurotransmitter, phenylethylamine (PEA), is tagged the 'love molecule' because it induces feelings of excitement and apprehension. PEA is found in chocolate and it, too, is linked to the feel-good effects of exercise. Overall, a swift jog could be more conducive to love than anything you might find in a bottle.

Love potions or not, it seems that drugs can carry both pros and cons when it comes to copulation.

❇ Please, bore me

In 1995, the internet was abuzz with news of a startling side effect of the antidepressant clomipramine. The effect was first described in a paper in the *Canadian Journal of Psychiatry* which stated that four patients taking the drug had spontaneous orgasms every time they yawned.

One of the patients, a woman, had been depressed for three months. The drug cured her problem, but she asked if she could be allowed to continue using it since she enjoyed the side effect so much. She had even found she could experience an orgasm by deliberate yawning. Another patient, a male, was also highly satisfied both with the drug's clinical effectiveness and its side effect. He asked to continue taking the drug, solving the 'awkward and embarrassing' problem of repeated spontaneous climaxes by wearing a condom all day.

Apparently some 5 per cent of users of clomipramine reported the side effect, which had also been observed in users of Prozac, although for the majority the ability to orgasm was inhibited by these drugs.

One writer suggested that the effect, if it became widespread, could have had interesting social consequences. People who experienced it would presumably seek out the most boring person they could find at parties ...

✳ Dopey sperm

In the year 2000, Herbert Schuel of the State University of New York at Buffalo claimed that cannabis may slow down sperm. This could explain reports of low fertility among men who smoked a lot of marijuana. Schuel's team treated sperm in a test tube with a synthetic anandamide – a chemical very similar to the active ingredient of cannabis. Low levels sent sperm swimming into overdrive but higher levels slowed their motion to a lazy crawl. The chemical also inhibited the ability of the sperm to bind to the egg and penetrate it. 'It really stops them cold,' Schuel said. Natural anandamides are present in semen, and are also secreted by the oviduct and egg follicles.

⚛ Frozen stiff

A frozen penis was the key to discovering the male impotence drug Viagra, according to three patent applications filed by Pfizer of Sandwich in Kent. In June 1990, Andrew Bell, David Brown and Nicholas Terrett described a range of pyrazolo-pyrimidinones that relieved angina and high blood pressure by inhibiting an enzyme called phosphodiesterase.

Three years later, Terrett and Peter Ellis tested the drug on a detached frozen human penis. They found it inhibited a similar enzyme in the corpus cavernosum tissue. When this tissue becomes engorged with blood it produces an erection. In a June 1997 patent, Peter Dunn and Albert Wood discovered a better way of making the drug which was to become Viagra.

⚛ Fart molecule could be next Viagra

The stink of flatulence and rotten eggs could provide a surprising lift for men. Hydrogen sulphide (H_2S) causes erections in rats and may one day provide an alternative to Viagra for men.

The penis is packed with spongy tissue that produces an erection when it fills with blood. Nitric oxide (NO) helps relax the walls of arteries that supply the penis, allowing extra blood to flow in. Viagra works by blocking an enzyme that destroys NO. In 2009, H_2S had been shown to relax the walls of major blood vessels too. Giuseppe Cirino at the University of Naples Federico II in Italy and his colleagues found enzymes that produce H_2S in human penile tissue. Injecting this tissue with H_2S dilated the blood vessels, while injecting it into the penises of live rats produced erections.

Of course, it's not size that matters, or so we are told.

✱ The lengths some men will go to

In 1995, it was discovered that penises were generally smaller than popularly assumed. By injecting 60 men with a drug that produces erections, and then measuring the size of their organs with a tape measure, Jack McAninch and Hunter Wessells of the University of California, San Francisco, concluded that an average erect penis is 12.8 centimetres long. (That's just over 5 inches for those who can't think about their anatomy in metric units).

Wessells and McAninch conducted their size survey after several unhappy patients came to them with complications resulting from penile augmentation operations. The researchers hoped that by revealing the 'norm' for penis size, they would help doctors to decide if a patient really needed such a procedure.

Only a few surgeons in the US performed penis-lengthening operations, often advertising aggressively in the sports pages of newspapers and charging up to $6000 for their services. But the results were less than perfect, said McAninch. The operation could result in uneven swelling, bleeding, loss of sensation and infections leading to skin loss and deformity. These could lead, in turn, to psychological problems, including the inability to achieve any erection at all. The researchers intended to firm up their data by measuring more penises.

There's no such thing as the ideal size or shape or body, we suspect. But drinking can help improve most things. Beauty is in the eye – or the beer glass – of the beholder.

✳ 'Beer goggles' are real – it's official

The next time you hear someone blaming 'beer goggles' for their behaviour, you may have to believe them. People really do appear more attractive when our perceptions are changed by drinking alcohol.

There have been few previous attempts to investigate the idea that people seem to find others more attractive when drunk. In 2003, psychologists at the University of Glasgow, UK, published a study in which they asked heterosexual students in campus bars and cafés whether they had been drinking, and then got them to rate photos of people for attractiveness. While the results supported the beer goggles theory, another explanation is that regular drinkers tend to have personality traits that mean they find people more attractive, whether or not they are under the influence of alcohol at the time.

To resolve the issue, a team of researchers led by Marcus Munafò at the University of Bristol in the UK conducted a controlled experiment. They randomly assigned 84 heterosexual students to consume either a non-alcoholic lime-flavoured drink or an alcoholic beverage with a similar flavour. The exact amount of alcohol varied according to the individual but was designed to have an effect equivalent to someone weighing 70 kilograms drinking 250 millilitres of wine – enough to make some students tipsy. After 15 minutes, the students were shown pictures of people their own age, from both sexes.

Both men and women who had consumed alcohol rated the faces as being more attractive than did the controls, and the effect was not limited to the opposite sex – volunteers who had drunk alcohol also rated people from their own sex as more attractive. This contrasted with the Glaswegian team's results, where there was only an effect when men were looking at pictures of women, and vice versa. One explanation, said Munafò, is that alcohol-boosted perceptions of attractiveness tend to become focused on potential sexual partners in environments conducive to sexual encounters. He aimed to repeat the experiment after showing students a video of people flirting in a bar, to provide some appropriate social cues.

As well as changing perceptions of attractiveness, alcohol also encourages us to engage in behaviour we would otherwise avoid. In a study by Robert Leeman of Yale University, students reported they were more likely to engage in risky sexual acts after drinking – which could be due to alcohol lowering our inhibitions through a direct effect on the brain or by providing a convenient excuse for such behaviour.

But if you've fitted your beer goggles yet no prospective partner heaves into view, don't switch on the football.

⚛ Shagged out

Here is a sobering message for footballers launching themselves into a new soccer season: an overactive sex life could give you a career-threatening injury. A study conducted in 1999 suggested that players were especially prone to arthritic knee injuries triggered by sexually transmitted bacteria.

The study, by Paul Oyudo, analysed the cases of ten sportsmen with persistent knee injuries. Of the ten, six were footballers and five of these played in the English Premier League. John King, a consultant orthopaedic surgeon at the London Independent Hospital and president of the British Sports Doctors Association, who released Oyudo's study, hoped to make footballers aware of the risk to their careers posed by frequent unprotected sex.

While most footballers' injuries were sustained on the field of play, the cases studied by Oyudo involved sexually acquired reactive arthritis, or SARA. It is triggered by the same bacteria that cause non-specific urethritis, an inflammation of the urethra. In eight of the ten cases reviewed by Oyudo, the sportsmen clearly had non-specific urethritis. Three had cloudy discharges in their urine and two tested positive for *Chlamydia trachomatis*, a bacterium implicated as a cause of SARA.

Whether the sportsmen revealed the full extent of their sexual activities to their doctors was unclear, but 5 of them admitted having had more than 11 sexual partners in their lifetime. Only about one in four British men in their mid-twenties reported having had this many partners. 'The level of promiscuity among these sportsmen calls for concern,' Oyudo wrote. 'Footballers appear to be the greatest culprits.'

A spokesman for the Football Association said at the time that young players were routinely given 'general sex education', including advice on avoiding sexually transmitted diseases.

Presumably this is not a problem for footballers in Hungary who, thanks to this device, will surely be keen to practise safe sex.

⚛ Ode to joy

Hungary made a special contribution to music in 1997 with the Serenading Condom, which used technology very similar to the kind that makes birthday and Christmas cards play a tinny tune when they are opened. A microswitch in the condom connected a small battery to a preprogrammed sound synthesiser and a tiny loudspeaker. When the condom was unfurled it played a tune.

Hungarian lovers could choose between two condoms, each with its own serenade. One played the traditional Hungarian tune *You Sweet Little Dumbbell*. The other sang the *Internationale*, also known as *Arise Ye Workers*.

⚛ The little shock that's too much for a sperm

In 1990, a contraceptive which worked by electrocuting sperm was developed in the US. It had already been successfully tested in baboons. The device, similar to a tiny heart pacemaker, consisted of a lithium/iodide battery – which was half a centimetre long and as thick as a cotton bud – and two electrodes. The plastic cylindrical battery was placed in the cervix and is anchored by two plastic lugs.

The 2.8-volt battery generated a constant electrical current of 50 microamps. This was conducted across the cervix by mucus or seminal fluid, immobilising sperm in three to four minutes, according to researchers at the Women's Medical Pavilion in New York who developed the device. They believed the current would prevent the sperm passing through the cervix and fertilising the ovum: their in vitro studies showed that 100 per cent of sperm were stopped in their tracks by this level of current, and studies in baboons seemed to back up the findings.

The batteries were a modified form of a pacemaker. However, while a pacemaker lasts for up to ten years, the contraceptive batteries would run out after a year.

Steven Kaali, medical director of the Women's Medical Pavilion, said human trials lasting years would be necessary before the device could be used. 'Everyone believes in their own invention – I think this is the best thing ever to happen to women, but the proof will come from the R&D stages,' said Kaali.

He believed that the device would have few side effects: there had been no reports of problems such as burns or chemical changes in people who wore pacemakers. An electrical current might also kill bacteria and fungi, and he hoped that contraceptives incorporating the battery could cut down the risk of sexually transmitted diseases.

Shock therapy during sex was probably never likely to be a winner. As useless as spraying your genitals with cola, perhaps?

✳ 'Coca-Cola douches' scoop Ig Nobel prize

Tests of whether drinks such as Coke and Pepsi could be used as spermicides were among the many offbeat ideas celebrated at the 2008 Ig Nobel awards. The tongue-in-cheek awards, presented at Harvard University, are organised by the humorous scientific journal the *Annals of Improbable Research* and awarded for research achievements 'that make people laugh – then think'.

Deborah Anderson of Harvard Medical School's birth-control laboratory took her first step towards the Ig Nobel chemistry prize in the 1980s when she asked medical student Sharee Umpierre what type of contraception had been used at the all-girl Catholic boarding school she had attended in Puerto Rico. 'Coca-Cola douches,' Umpierre replied.

'Coca-Cola douches had become a part of contraceptive folklore during the 1950s and 1960s, when other birth-control methods were hard to come by,' Anderson told *New Scientist*. 'It was believed that the carbonic acid in Coke killed sperm, and the method came with its own "shake and shoot applicator" – the classic Coke bottle.'

To see if Coke really worked, Anderson, Umpierre and their colleague, gynaecologist Joe Hill, mixed four different types of Coke with sperm in test tubes. A minute later, all sperm were dead in the Diet Coke, but 41 per cent were still swimming in the just-introduced New Coke. But that's not good enough, Anderson warned. Sperm 'can make it into the cervical canal, out of reach of any douching solution, in seconds' – faster than anyone could shake and apply a bottle of Diet Coke.

The three researchers shared the chemistry prize with Chuang-Ye Hong of the Taipei Medical University in Taiwan and his colleagues, whose similar experiment found both Coca-Cola and its arch-rival Pepsi-Cola useless as spermicides.

But, even with cola as protection, you still need to be very careful whom you choose to kiss.

✳ Read my lips

In 1993, the journal *Nature* suggested that the phrase 'kiss and tell' might take on a new meaning when it reported that after humans have done the kissing, bacteria and fungi do the telling. The article alleged that each of us has a different 'lip fauna', as individual as our fingerprints. When people kiss, there is a veritable battle of the bacteria as the two microscopic populations come into contact.

Close family members who touch frequently share many of the same skin bacteria, but a stranger has a totally different fauna.

The 'locals' eventually come out on top, as they are better suited to their environment. So, by comparing the ratios of 'foreign' and 'native' microbes remaining, scientists could not only tell who had been kissing whom but also when they kissed.

The techniques were devised for forensic work in legal cases involving bodily contact. But with the number of high-profile personages implicated in extramarital indiscretions, how long might it be before forensic scientists are called in to read their lips?

Scientists, we note with some pride, are prepared to ask the questions others only dare think about, or indeed do not even consider at all. Otherwise, how would we know that, instead of stunting your growth, this almost universal practice might actually be good for you …

❄ Masturbating may protect against prostate cancer

It will make you go blind. It will make your palms grow hairy. Such myths about masturbation are largely a thing of the past. But research conducted in 2003 had even better news for young men: frequent self-pleasuring could protect against the most common kind of cancer.

A team in Australia led by Graham Giles of The Cancer Council Victoria in Melbourne asked 1079 men with prostate cancer to fill in a questionnaire detailing their sexual habits, and compared their responses with those of 1259 healthy men of the same age. The team concluded that the more men ejaculated between the ages of 20 and 50, the less likely they were to develop prostate cancer. The protective effect was greatest while men were in their twenties: those who had ejaculated more than five times per week in their twenties, for instance, were one-third less likely to develop aggressive prostate cancer later in life.

The results contradicted those of previous studies, which suggested that having had many sexual partners, or a high frequency of sexual activity, increased the risk of prostate cancer by up to 40 per cent. The key difference was that these earlier studies defined sexual activity as sexual intercourse, whereas the more recent study focused on the number of ejaculations, whether or not intercourse was involved.

The team speculated that infections caused by intercourse may increase the risk of prostate cancer. 'Had we been able to remove ejaculations associated with sexual intercourse, there should have been an even stronger protective effect of other ejaculations,' they suggested. 'Men have many ways of using their prostate which do not involve women or other men,' Giles added.

But why should ejaculating more often cut the risk of prostate cancer? The team speculated that ejaculation prevents carcinogens building up in the gland. The prostate, together with the seminal vesicles, secretes the bulk of the fluid in semen, which is rich in substances such as potassium, zinc, fructose and citric acid. Generating the fluid involves concentrating these components from the bloodstream up to 600-fold – and this could be where the trouble starts. Studies in dogs showed that carcinogens such as 3-methylcholanthrene, found in cigarette smoke, are also concentrated in prostate fluid. 'It's a prostatic stagnation hypothesis,' said Giles. 'The more you flush the ducts out, the less there is to hang around and damage the cells that line them.'

'All these mechanisms are totally speculative,' cautioned breast cancer expert Loren Lipworth of the International Epidemiology Institute in Rockville, Maryland.

But if the findings were ever confirmed, future health advice from doctors might no longer be restricted to diet and exercise. 'Masturbation is part of people's sexual repertoire,' said Anthony Smith, deputy director of the Australian Research Centre in Sex, Health and Society at La Trobe University in Melbourne. 'If these findings hold up, then it's perfectly reasonable that men should be encouraged to masturbate,' he said.

We later discovered that sex generally may help prevent prostate cancer.

✣ Frequent ejaculation may protect against cancer

Frequent sexual intercourse and masturbation could protect men against a common form of cancer, suggested a 2004 study of the issue. The US study, which followed nearly 30,000 men over eight years, showed that those who ejaculated most frequently were significantly less likely to get prostate cancer. The results backed the findings of the Australian study cited above which asserted that masturbation was good for men.

In the US study, the group with the highest lifetime average of ejaculation – 21 times per month – were a third less likely to develop the cancer than the reference group, who ejaculated four to seven times a month. Michael Leitzmann, at the National Cancer Institute in Bethesda, Maryland, and colleagues set out to test a long-held theory that suggested the opposite – that a higher ejaculation rate raises the risk of prostate cancer. 'The good news is it is not related to an increased risk,' he said. In fact, it 'may be associated with a lower risk'.

'It goes a long way to confirm the findings from our recent case-control study,' said Graham Giles, who led the Australian study. He praised the study's large size – including about

1500 cases of prostate cancer. Furthermore, it was the first study to begin by following thousands of healthy men. This ruled out some of the biases which might be introduced by asking men diagnosed with prostate cancer to recall their sexual behaviour retrospectively.

At the start of the study, the men filled in a history of their ejaculation frequency and then filled in further question-naires every two years. Men of different ages varied in how often they ejaculated, so the team used a lifetime average for comparisons. Compared with the reference group who ejaculated four to seven times a month, 'each increase of three ejaculations per week was associated with a 15 per cent decrease in the risk of prostate cancer', said Leitzmann.

'More than twelve ejaculations per month would start conferring the benefit – on average every second day or so,' he said. However, whilst the findings were statistically significant, Leitzmann remained cautious. 'I don't believe at this point our research would warrant suggesting men should alter their sexual behaviour in order to modify their risk.'

A further caveat was that the benefit of ejaculation was less clear in relation to the most dangerous, metastasising form of prostate cancer, compared with the organ-confined or slow-growing types.

Giles noted that neither study examined ejaculation during the teenage years – which may be a crucial factor. But he said: 'Although much more research remains to be done, the take-home message is that ejaculation is not harmful, and very probably protective of prostatic health – and it feels good!'

Sadly, research released in 2009 seems to have shown that the added protection only occurs if frequent ejaculation takes place in those aged 40 or over, which rather confuses the issue. Indeed, as was once thought, it is possible that prostate cancer may be more common in men who are sexually active at a young age. There was always bound to be a downside wasn't there? Clearly, more research needs to be done to ascertain the true picture. All we can do is present the evidence and leave it in readers' hands ... And there's something else you should think twice about doing too.

✳ Oral sex linked to mouth cancer

Oral sex can lead to oral tumours. That was the conclusion of researchers who, in 2004, proved what had long been suspected – that the human papilloma virus can cause oral cancers.

The risk, thankfully, is tiny. Only around 1 in 10,000 people develop oral tumours each year, and most cases are probably caused by two other popular recreational pursuits: smoking and drinking. The researchers were not recommending any changes in behaviour.

The human papilloma virus (HPV), an extremely common sexually transmitted infection, had long been known to cause cervical cancers. Several small studies suggested it also plays a role in other cancers, including oral and anal cancers.

'There has been tremendous interest for years on whether it has a role in other cancers. Many people were sceptical,' said Raphael Viscidi, a virologist at Johns Hopkins University School of Medicine in Baltimore, Maryland, a member of the team which did the latest work.

The researchers, working for the International Agency for Research on Cancer in Lyon, France, compared 1670 participants who had oral cancer with 1732 healthy volunteers. The patients lived in Europe, Canada, Australia, Cuba and Sudan. HPV16, the strain seen most commonly in cervical cancer, was found in most of the oral cancers too. The people with oral cancers containing the HPV16 strain were three times as likely to report having had oral sex as those whose tumour did not contain HPV16. There was no difference between men and women in terms of how likely the virus was to be present in the cancers. The researchers thought both cunnilingus and fellatio could infect people's mouths.

Patients with mouth cancer were also three times as likely to have antibodies against HPV as the healthy controls. For cancers of the back of the mouth, the link was even stronger.

Cancer specialist Newell Johnson of King's College London agreed. 'We have known for some time that there is a small but significant group of people with oral cancer whose disease cannot be blamed on decades of smoking and drinking, because they're too young,' he said. 'In this group there must be another factor, and HPV and oral sex seems to be one likely explanation. This study provides the strongest evidence yet that this is the case.'

Genital HPV infections are common. At any one time, around a third of 25-year-old women in the US are infected. It is thought that only 10 per cent of infections involve cancer-causing strains, and that 95 per cent of women will get rid of the infection within a year. But even this does not explain why so few develop cancer.

And if you thought you'd had enough bad news, it turns out that the old adage about the bad guys and their girls really is true.

Bad guys really do get the most girls

Nice guys already knew it, and in 2008 two studies confirmed it: bad boys get the most girls. The finding may help explain why a nasty suite of antisocial personality traits known as the 'dark triad' persists in the human population, despite their potentially grave cultural costs.

The traits are the self-obsession of narcissism; the impulsive, thrill-seeking and callous behaviour of psychopaths;

and the deceitful and exploitative nature of Machiavellianism. At their extreme, these traits would be highly detrimental for life in traditional human societies. People with these personalities risk being shunned by others and shut out of relationships, leaving them without a mate, hungry and vulnerable to predators. But being just slightly evil could have an upside: a prolific sex life, according to Peter Jonason at New Mexico State University in Las Cruces. 'We have some evidence that the three traits are really the same thing and may represent a successful evolutionary strategy.'

Jonason and his colleagues subjected 200 college students to personality tests designed to rank them for each of the dark triad traits. They also asked about their attitudes to sexual relationships and about their sex lives, including how many partners they'd had and whether they were seeking brief affairs. High 'dark triad' scorers are more likely to try to poach other people's partners for a brief affair. The study found that those who scored higher on the dark triad personality traits tended to have more partners and a greater desire for short-term relationships. But the correlation only held in males.

James Bond epitomises this set of traits, Jonason said. 'He's clearly disagreeable, very extroverted and likes trying new things – killing people, new women.' Just as Bond seduces woman after woman, people with dark triad traits may be more successful with a quantity-style or shotgun approach to reproduction, even if they don't stick around for parenting. 'The strategy seems to have worked. We still have these traits,' Jonason said.

This observation seemed to hold across cultures.

But questions remained as to its effectiveness. 'They still have to explain why it hasn't spread to everyone,' said Matthew Keller of the University of Colorado in Boulder. 'There must be some cost of the traits.' One possibility, both Keller and Jonason suggested, is that the strategy is most successful when dark triad personalities are rare. Otherwise, others would become more wary and guarded.

But, finally, take comfort in the fact that bizarre sexuality is not confined to humans.

✳ Female beetles have a thirst for sex

Buying a lady a drink to win her favour is a trick not confined to men. Some beetle females will mate simply to quench their thirst.

The bean weevil *Callosobruchus maculatus* feeds on dry pulses. With a diet like this, the male's ejaculate is a valuable water source for females. In 2007, Martin Edvardsson at Uppsala University, Sweden, tested the idea that females tap into this by keeping them on dry beans with or without access to water. Females living on beans alone accepted more matings, presumably to secure the water in the seminal fluid.

Edvardsson said that the energy used to produce the ejaculate, which makes up a whopping 10 per cent of a male's weight, is well spent. Once impregnated, females lose interest in sex – probably to avoid further injury from the male's spiny penis. They are more likely to mate again if they are thirsty. 'This is a massive investment for the male,' Edvardsson said. 'It buys them time before the females remate and their sperm have to compete with that of other males.'

Females with access to water lived on average for a day and a half longer than those without water. Since average lifespan is only around nine days, this makes quite a difference to the total number of eggs they can lay.

⚛ Turkey turn-ons

While researching the sexual behaviour of turkeys in the 1960s, Martin Schein and Edgar Hale of Pennsylvania State University discovered that male members of that species truly are not fussy. When placed in a room with a lifelike model of a female turkey, the birds mated with it as eagerly as they would the real thing.

Intrigued by this observation, Schein and Hale embarked on a series of experiments to determine the minimum stimulus it takes to excite a male turkey. This involved removing parts from the turkey model one by one until the male bird eventually lost interest. Tail, feet and wings – Schein and Hale removed them all, but still the clueless bird waddled up to the model, let out an amorous gobble, and tried to do his thing. Finally, only a head on a stick remained. The male turkey was still keen. In fact, it preferred a head on a stick to a headless body.

The researchers speculated that the males' head fixation stemmed from the mechanics of turkey mating. When a male turkey mounts a female, he is so much larger than her that he covers her completely, except for her head. Therefore, they suggested, it is her head that serves as his focus of erotic attention.

Schein and Hale then went on to investigate how minimal they could make the head before it failed to excite the turkey. They discovered that a freshly severed head on a stick worked best. Next in order of preference was a dried-out male head, followed by a two-year-old 'discolored, withered, and hard' female head. Last place went to a plain balsa wood head, but even that elicited a sexual response. They published their results in 1965 in a book called *Sex and Behavior*.

Before we humans snigger at the sexual predilections of turkeys, we should remember that our species stands at the summit of the bestial pyramid of the perverse. Humans will attempt to mate with almost anything. A case in point is

Thomas Granger, the teenage boy who in 1642 became one of the first people to be executed in puritan New England. His crime? He had sex with a turkey.

✳ Double the fun

Male lizards and snakes have to make a choice when they mate: which penis should they use? They have two. One is connected to their right testis, and one to their left. In 1990, two biologists from the US found that some lizards alternate their penises when each is depleted of sperm. In this way, they maximise their chance of fertilising a female. Richard Torkaz and Joseph Slowinski of Miami University studied the lizard *Anolis carolinensis*. They found that the penis a lizard used depended on which they had used last time they mated, and how long ago they had last used it.

Torkaz and Slowinski believed that such behaviour could be explained by 'sperm competition'. Zoologists have realised that competition for a female's favour does not simply involve fights among males or displays of bright plumage; it can continue inside the female's reproductive tract as sperms compete to be first to the egg.

In response to such sperm competition, males of some species developed sperm that formed into a hard plug, blocking the way for the sperm of any other males. In other species, males produced a lot of sperm, so that these outnumbered and swamped the opposition. However, keeping up a high level of sperm production can be a problem. If successive copulations followed on too quickly from one another, this could result in the number of sperm in the ejaculate dropping sharply. By alternating their penises, lizards of *A. carolinensis* may overcome this problem. Torkaz and Slowinski found that males alternated the use of their right and left penises only if one copulation followed less than 24 hours after another. If the interval between copulations was longer, then they used the same penis.

The biologists carried out an experiment in which they prevented penis alternation. They did this by taping over one half of a male's cloacal vent, through which the penis protrudes when aroused. This stopped the animal from using the penis on that side. The biologists found that males which were forced to reuse the same penis within 24 hours transferred fewer sperm in the later copulations. However, the researchers found no such drop in the numbers of sperm when successive copulations were as long as 72 hours apart. In this situation, they said, the sperm duct had time to be restocked with sperm.

According to Torkaz and Slowinski, their research demonstrated that the lizards adjusted their behaviour in response to sperm competition, using their penises alternately to ensure that maximum sperm transfer was kept up, even if the gap between copulations was short.

⚛ Big attraction

Few birds have penises, but the Argentine lake duck *Oxyura vittata* is one sizeable exception. Its corkscrew-shaped member was thought to be about 20 centimetres long – half the duck's body length. But in 2001, Kevin McCracken of the University of Alaska found one aroused specimen whose penis was hanging out of the body sac where it usually resided. It measured a whopping 42.5 centimetres.

'This is the first record of the penis "hanging out" in its natural form,' said McCracken. He suggested that this highly promiscuous bird displayed its penis to attract females, which might explain why evolution has favoured length.

And, after apologising for going to such lengths, we move on to the next, more genteel, chapter.

7 Animals and their ilk

'If we could talk to the animals, learn their languages, maybe take an animal degree …' Dr Doolittle, in the eponymous hit musical, mused what it would be like if we could speak to those with which humans share the planet. Sadly, we are not much further down the road to deciphering the language of intelligent dolphins or chimps, let alone actually figuring out if dogs bark in order to converse or simply to irritate the neighbours.

In fact, some may feel that science has had a patchy relationship with the animal kingdom, but *New Scientist* has long campaigned for an end to the sight of mammals with cosmetics rubbed in their eyes and smoking beagles. Indeed, often it's the proverbial Mother Nature rather than humans who can be the wickedest of mistresses. What personality disorder led her to decree that herrings should communicate via farts? More intriguingly, what on earth led researcher Ben Wilson and his colleagues to uncover this fact? And why did Mother Nature decide that male rats should sing an ultrasonic song shortly after ejaculating? Just so biologists Ronald Barfield and Lynette Geyer could discover it and find their way into this chapter?

From classical-music-loving carp to urban myths about toppling penguins, little has escaped the notice of scientists down the years, although sadly we had to leave out the story of the Kenyan MP who suggested allowing hyenas into the nation's hospitals to consume the unclaimed bodies of deceased humans. He thought it would save money on storage and burials (and who could argue with such logic?). Unfortunately (for us, not the relatives of the cadavers), he doesn't qualify for this chapter because a) his suggestion was inexplicably rejected and b) he was a politician, not a scientist, which really disqualifies him from any knowledge of what's good for any of us, let alone an appearance in this book.

And we admit it, this chapter is just as much about animals pushing the boundaries of what is believable or acceptable as it is about scientists. But we all share a common ancestor, so what the heck? French poet Jean Marcenac almost wrote 'The more I learn about humans, the more I like animals'. We'd only agree up to a point, especially when faced with a farting herring ...

While most of the experiments on animals that follow are based on passive research, we are surprised but pleased to include this report that shows spiders are quite chilled when it comes to heavy drug consumption. Elephants on LSD was a particularly nasty idea (see page 130), but, it seems, spiders are rather more groovy.

✷ Spiders on speed get weaving

Spiders on marijuana are so laid back that they weave just so much of their webs and then ... well, it just doesn't seem to matter anymore. On the soporific drug chloral hydrate, they drop off before they even get started.

A spider's skill at spinning its web is so obviously affected by the ups and downs of different drugs that in 1995 scientists at NASA's Marshall Space Flight Center in Alabama thought spiders could replace other animals in testing the toxicity of chemicals.

Different drugs had varying effects on the average arachnid addict. On benzedrine, a well-known upper, house spiders spun their webs with great gusto, but apparently without much planning, leaving large holes. On caffeine they seemed unable to do more than string a few threads together at random.

The more toxic the chemical, the more deformed the web. NASA researchers had hoped that with help from a computer program they would be able to quantify this effect to produce an accurate test for toxicity.

But, being terribly right-on, we can't really say we are happy with dosing spiders or any animal with drugs already tried and tested in humans. Far better, we feel, to give them something we can presume in advance they might enjoy.

Animal pleasure

In 1995, primate pornography was the entertainment on offer for bored gorillas at Longleat House near Warminster, UK. Samba and Nico romped over a quarter of a hectare of private island during the day and enjoyed all the comforts of home at night.

Their house was equipped with satellite TV, to which they were glued throughout the long winter evenings. Longleat's press officer Claire Keener told *New Scientist* that Samba and Nico preferred jungle movies and became especially excited when gorillas appeared on the screen. At the consenting age of 30 years old, both were allowed to watch blue movies of primates mating in the wild, in the hope that they would be aroused into mating.

Not a bad life: first-class room service, a wholesome and varied diet of fruit, vegetables and beech leaves, and a visit from a personal vet once a week. Lights out at bedtime ensured a comfortable night's sleep but despite the movies, there was no monkeying around.

Success, sadly, was not forthcoming, which is bad news for all those zoological institutions that have trouble getting the likes of pandas and polar bears to mate. Still, panda porn and its ilk give us an excuse to report any amount of research into the sexual behaviour of animals.

Exhibitionist spiny anteater reveals bizarre penis

In 2007, the bizarre sex life of the spiny anteater was exposed when it was discovered that the male ejaculates using only one half of its penis. These findings about the creature's sex life may seem salacious but they could help shed light on an evolutionary mystery.

It seems that the way the mammal ejaculates is similar to the way reptiles do – by shutting down one side of its penis before secreting semen from the other side. Reptiles have a pair of male members called hemipenes for sex, and they use only one of the two during each act of copulation (see page 167).

The spiny anteater (*Tachyglossus aculeatus*), also known as the short-beaked echidna, is a primitive mammal found in Australia and New Guinea. Like the platypus, it is a monotreme, laying eggs instead of bearing live young. Monotremes have many features in common with reptiles, and the hope is that by studying them, scientists may find clues as to how mammals evolved. The spiny anteater, however, is notoriously difficult to observe in the wild and shows little enthusiasm for breeding in captivity, so, prior to 2007, nobody had managed to observe them ejaculate.

Fortunately, Steve Johnston of the University of Queensland in Gatton, Australia, and his colleagues inherited a male spiny anteater that was not so shy. The creature had been 'retired' from a zoo because it produced an erection when being handled at public viewing sessions, bemusing its visitors. By filming this animal, the researchers were able to describe the unique spiny anteater erection and ejaculation behaviour for the first time.

The spiny anteater's four-headed phallus had been puzzling scientists. 'When we tried to collect semen by electrically stimulated ejaculation before, not only did we not get a single drop, but the whole penis swelled up to a four-headed monster that wouldn't fit the female reproductive tract, which has only two branches,' said Johnston. 'Now we know that during a normal erection, two heads get shut down and the other two fit.' The heads used are swapped each time the mammal has sex.

The evolutionary significance of one-sided ejaculation was unknown, but may play a role in sperm competition – where sperm from many males may compete to fertilise an egg. Indeed, in the spiny anteater, up to 11 males may form a queue behind one female to copulate with her. The researchers also observed that hundreds of sperm team up to form bundles that swim much faster than individual sperm in the spiny anteater's semen – another possible adaptation for sperm competition.

'We can now study echidna sperm much better, which should offer fascinating insights into the evolution of mammals,' said Russell Jones from the University of Newcastle in New South Wales, who first dissected sperm bundles from dead echidna in the 1980s.

✵ Why a big horn gives beetles a tiny organ

Rapid metamorphosis may explain why beetle species are numerous enough to overwhelm almost any other creature on Earth. The superfast evolution of the male dung beetle's penis to wildly differing sizes in the same species could mean that new species appear within a few years rather than over millennia.

In 2008, Armin Moczek from Indiana University, Bloomington, and colleagues studied four populations of one species of horned beetles in Eastern and Western Australia, Italy and the US. The most striking difference was between beetles in Western Australia, which had small horns and big genitals, and those in the US, which had the complete opposite. The variation in genital size between these two populations was as pronounced as that between ten other species in countries across Asia, Europe, and South America.

Organs developing within the pupa must compete for a limited supply of nutrients. So if one organ grows bigger, another is stunted. Thus, males with large horns tend to have smaller genitals, and vice versa.

Strong male beetles use their horns to fight for females, but weaker males prefer to sneak off to mate while competitors are fighting. Which strategy works best depends on the size of the population. In those with more females, fighting is most successful, and big-horned beetles win most mates. But when there are not enough females, fighting is often fruitless, so evolution favours beetles with smaller horns but cunning tactics.

Because beetles with genitals of different sizes cannot mate, Moczek thought the four populations might soon split into distinct species. 'It's unprecedented that 50 years would be long enough to generate variation normally only found in species which have been separated for millions of years,' he said.

Of course, mating practices vary greatly between species, but it's safe to assume the medaka fish in this following article from 1994 weren't expecting to breed in zero gravity.

✸ Astronauts aim to catch fish in the act

Voyeurism will reach new heights this week as astronauts aboard the space shuttle *Columbia* video the mating behaviour of two pairs of Japanese medaka fish. If the fish succeed in mating in orbit, and produce eggs and offspring, scientists will have a unique chance to study the role gravity plays in the development of the embryo. Successful mating could also pave the way for further experiments aimed at providing astronauts on long missions with a renewable source of food.

The experiment is one of about 80 set up by an international team of scientists to fly in a microgravity laboratory aboard *Columbia*. In the absence of gravity, many species of fish lose their balance and swim in loops rather than straight lines. This makes mating unlikely, says Victor Schneider, one of the NASA scientists overseeing the programme. Japan's national space agency, NASDA, took a number of species up in aircraft flying a parabolic course that reproduces weightless conditions for a few seconds. The small, orange-coloured medakas did best in the flight tests, hence their spaceflight. The species also has other advantages. Medakas' eggs take only eight days to hatch, so the astronauts should see results before the end of the 14-day mission. In addition, fish, eggs and offspring are all transparent, so the astronauts will be able to record the development of the body and internal organs on film. After behaving strangely at first, the fish recovered, but mating was a tentative affair.

None of us is surprised that animals enjoy sex, of course. What is more perturbing is that, like humans, they might have hang-ups about it.

✳ The cat's whiskers

Many inventors claim that their latest idea is 'the dog's bollocks'. But in 1996, CTI Corporation of Buckner, Missouri could claim that theirs really was. The company had developed a line in polypropylene canine testicles.

Artificial testicles, according to *Chemical and Engineering News*, were selling like hot cakes in the US and Canada. Trade-named Neuticles, they were installed in the dog's scrotum in a two-minute procedure immediately following removal of the original articles. The idea was to be able to neuter dogs without causing psychological trauma.

Gregg Miller, CTI's president, suggested that Neuticles would encourage owners to have their dogs seen to: 'With these, the dog looks the same. He feels the same. He doesn't even know he's been neutered.'

Although Miller admitted that some people thought the product was silly, he claimed that Neuticles were 'big news in the veterinary industry'. He had even produced 'I love Neuticles' bumper stickers for the proud owners of dogs with ersatz balls. Veterinarians, however, were sceptical of the idea that Neuticles could do much for a dog's self-esteem.

And so are we. Even so, could this Suffolk punch stallion have benefited from a pair? Probably not. Still, we are impressed by this example of school science in action.

✳ Soft solution

Childhood memories of a school science experiment helped a vet to save a patient's professional career in 1992. The unfortunate victim of an accident at work was a Suffolk punch stallion, kicked in the genitals by a mare he was supposed to service. A painful swelling developed on the stallion's penis.

The next day, vet Philip Ryder-Davis was called in. 'Despite all manner of treatment, the oedematous swelling got worse and worse over the next three days,' he said. With disaster looming, a long-forgotten science lesson came to mind, he explained.

Ryder-Davis remembered an experiment to demonstrate osmosis. The shell was removed from a raw egg, and the egg with its inner membrane still intact was put in a strong solution of sugar. Osmotic pressure forced the water from the egg, as it attempted to dilute the sugar solution and produce equilibrium on both sides of the membrane, and the egg gradually began to shrink. Ryder-Davis suspended a canister of syrup from belts around the animal's abdomen and placed the damaged organ inside. Within 24 hours the swelling had disappeared, he reported, and the stallion recovered.

A specialist in the biology of membranes confirmed that the treatment was feasible. 'Putting a swollen organ in a highly concentrated solution of salt or sugar would certainly pull the water out. It raises the question of what happens to the penises of swimmers in the Dead Sea,' he added.

Animals tend to avoid the Dead Sea – the hint is in the name. But, then again, you'd also expect animals to avoid falling into vessels from which they cannot escape. But they don't, which leads us nicely on to a very eccentric invention.

✳ Spidey's escape

It is an old adage of the patents profession that the oddest ideas are the easiest to patent, simply because nobody has ever tried before. Although the examiners searched a wide range of previous patents, they were not able to come up with anything that anticipated the application filed by Edward Doughney of Bedfordshire, UK, in 1994.

Spiders are often trapped inside a bath because they cannot climb up the slippery curved slope. So the bather either has to flush them away down the plughole or give them a helping hand. A more humane answer, Doughney suggested, was a spider ladder.

A strip of flexible plastic had a suction pad on one end which anchored it to the top of the bath. The strip had a rough surface, so an athletic spider could climb to safety unaided.

Of course, there's a whole raft of inventions designed to improve the lot of the animal kingdom. We've already encountered Neuticles and spider ladders, now how about these:

✳ Wonder baas

English Nature decided that old methods were the best when in 1995 the rural heritage authority wanted to clear a thorny patch of ground to encourage the growth of sea-wort, a favourite food of the endangered reddish buff moth's larvae. So a herd of milking goats was brought in to do the job on the Isle of Wight.

Everything went fine, with the goats munching happily on the scrub, clearing it so that full sunlight could reach the underlying sea-wort. But a problem emerged: the goats' udders and teats started getting badly scratched by low-lying thistles and thorns. 'It was spoiling the goats' milking potential, so we had to do something,' said David Sheppard of the English Nature Species Recovery Programme.

The solution was a bag made of tough nylon which slipped under the udder and strapped round the goat's back. English Nature claimed a world first – unless somebody else had already invented the thorn-proof bra.

Bleeping Miss Daisy

In 1993, according to *New Internationalist*, cowherds in Japan were in short supply, so researchers tried developing pagers to attach to cows' collars. They could then be 'bleeped' at milking time. The cows responded to musical notes, and were apparently particularly attracted by piano melodies. Only two weeks were needed to train the cattle to come home when paged.

For farmers who were unwilling to invest in this technology, there was a much cheaper method – simply playing the appropriate piano tunes over loudspeakers at milking time. And to keep on playing until the cows came home.

Cows like music. So, it seems, do carp:

An ear for baroque

In 2008, Ava Chase of the Rowland Institute at Harvard in Cambridge, Massachusetts, showed that carp could tell the difference between baroque music and John Lee Hooker, depressing a button with their snouts to indicate which was which. Carp do not even use sound to communicate, but they are renowned for their sensitive hearing. Similarly,

Java sparrows could not only distinguish between Bach and Schoenberg but could also, according to the findings of Shigeru Watanabe at Keio University in Tokyo, Japan, apply what they had learned about the differences between classical music and the more modern stuff to discriminate between the beautiful melodies of Antonio Vivaldi and the more atonal strains of Elliott Carter. Watanabe's sparrows also appeared to engage with the music, showing clear preferences for the prettier, more harmonious excerpts and choosing to listen to these rather than sit in silence.

… and hens like something else.

⚛ Hen night

A 1958 report from Germany threw an unexpected sidelight on the behaviour of hens. The agricultural research institute at Würzburg had been studying what, if any, were the harmful effects of wines produced from hybrid grapes. Germany, like France and Switzerland, was plagued with a sizeable output of wine of extremely low quality and dubious potability. In the experiments hens, spiders and goldfish were all used, and the hens in particular consumed astounding quantities of wine.

They were divided into three groups, the controls being given water only and the others pure wine and hybrid wine respectively. Each hen was given half a pint of liquid a day. Incredibly, 16 hens got through 600 pints of red wine in four months, seemingly in radiant health. These 16 did not, however, include any of the hens that drank hybrid wine – these died early in the study. The fate of the spiders and gold-fish was not disclosed.

We hope this experiment would be deemed illegal today, and we hope the hens died happy. Observing animal behaviour, of course, is nothing new. But some researchers notice the strangest things, leaving you wondering how on earth they came to be looking for that!

✵ Not tonight

While making a 'standard observation of sexual behaviour' in 1972, Ronald Barfield and Lynette Geyer noticed that the male of the species under study sang an ultrasonic song after his ejaculation. Following up this chance observation, the Rutgers University biologists found that this post-ejaculatory ultrasonic chant – which appeared to correlate with the individual's contented stertorous breathing – was common to all the males they studied. The duration of the song, and the behaviour of the female partners during its performance, suggested that it functioned as a desist-contact – or 'leave me alone' – signal, while the male recovered for another bout of sexplay.

The species involved in the experiments was the rat, whose sexual behaviour, according to the authors, had been 'exhaustively studied'. A series of mount bouts led to intromission and subsequent ejaculation, followed by a refractory period during which the male recuperated. The female hopped, darted about and wiggled her ears during copulation, but in the male's refractory period – when he usually either stretched out for a doze on the floor or groomed quietly – she refrained from such provocative behaviour.

Barfield and Geyer showed that for at least three-quarters of the interval between copulatory bouts the male emitted a series of calls at 22 kilohertz, apparently arising from the long exhalations of its languid breathing. Since the length of this calling period coincided neatly with the time during which the male was quite incapable of performing again, and since the behaviour of the female changed during this period too, the researchers suggested that the call conveyed a positive 'I'm out of action' type of message to the female.

Their conclusion was reinforced by the fact that a 22-kilohertz call was also characteristic of males that have been roundly defeated in a fight, and in females who were attempting to resist the mounts of overly attentive males. Perhaps, Barfield and Geyer suggested, the call generally reflected a state of social withdrawal; that '22 kilohertz is a basic "carrier frequency" for signals denoting states of contact avoidance'. It would be interesting to know what would happen if a rat colony was played continuous recordings of these antisocial signals. Could one devise an ultrasonic rat contraceptive?

We're not sure we want to find out. Nor are we sure how interested we are in fish farts, but somebody got out there and did the research ...

⚛ Fish farting may not just be hot air

In 2003, biologists linked a mysterious underwater farting sound to bubbles coming out of a herring's anus. No fish had been known before to emit sound from its anus or to be capable of producing such a high-pitched noise. 'It sounds just like a high-pitched raspberry,' said Ben Wilson of the University of British Columbia in Vancouver, Canada. Wilson and his colleagues could not be sure why herring made this sound, but initial research suggested that it might explain the puzzle of how shoals keep together after dark.

'Surprising and interesting' was how aquatic acoustic specialist Dennis Higgs, of the University of Windsor in Ontario, described the discovery. It was the first case of a fish potentially using high frequency for communication, he believed. Arthur Popper, an aquatic bio-acoustic specialist at the University of Maryland, was also intrigued. 'I'd not have thought of it, but fish do very strange and diverse things,' he said.

Fish are known to call out to potential mates with low 'grunts and buzzes', produced by wobbling a balloon of air called the swim bladder located in the abdomen. The swim bladder inflates and deflates to adjust the fish's buoyancy. The biologists initially assumed that the swim bladder was also producing the high-pitched sound they had detected. But then they noticed that a stream of bubbles expelled from the fish's anus corresponded exactly with the timing of the noise. So a more likely cause was air escaping from the swim bladder through the anus.

It was at this point that the team named the noise Fast Repetitive Tick (FRT). But Wilson pointed out that, unlike a human fart, the sounds were probably not caused by digestive gases because the number of sounds did not change when the fish were fed. The researchers also tested whether the fish were farting from fear, perhaps to sound an alarm. But when they exposed fish to a shark scent, there was again no change in the number of FRTs.

Three observations persuaded the researchers that the FRT was most likely produced for communication. Firstly, when more herring were in a tank, the researchers recorded more FRTs per fish. Secondly, the herring were only noisy after dark, indicating that the sounds might allow the fish to locate one another when they could not be seen. Thirdly, the biologists knew that herring could hear sounds of this frequency, while most fish cannot. This would allow them to communicate by FRT without alerting predators to their presence.

Wilson emphasised that this idea was just a theory. But the discovery is still useful, he said. Herring might one day be tracked by their FRTs, in the same way that whales and dolphins are monitored by their high-pitched squeals.

Modern technology and animals often don't mix. The urban legend that follows could so easily have been true and, while amusing, would have been irritating for the birds. So we are glad to put it to rest.

✳ No pushover

It's official: penguins don't fall over backwards when aircraft fly overhead. British pilots came back from the 1982 Falklands War with stories of penguins toppling over. Concerned that increasing air traffic might endanger wildlife, a team led by Richard Stone of the British Antarctic Survey spent five weeks in 2001 watching a thousand king penguins on South Georgia. After numerous overflights by two Royal Navy Lynx helicopters, 'Not one king penguin fell over,' Stone told Reuters.

We suspect the following is an urban legend too, but if it isn't, it's a classic amalgam of student tomfoolery and spectacular eccentric research.

✳ Landing strip

And here is another possibly apocryphal story sent to us by a friend who got it from a friend who … etc. It concerns a student at the Massachusetts Institute of Technology who went to the Harvard football ground every day for an entire summer wearing a black and white striped shirt. He would walk up and down the pitch for 10 to 15 minutes throwing birdseed all around him, blow a whistle and then walk off the field. At the end of the summer, the Harvard football team played its first home match to a packed crowd. When the referee walked on in his black and white strip and blew his whistle, hundreds of birds descended on the field and the game had to be delayed for half an hour while they were removed.

The student, so the story goes, wrote his thesis on this, and graduated.

It seems you can teach animals a great deal. Wielding technology they are even helping in the fight against crime.

⚛ Gerbils crack down

In 1982, drug smugglers at Canadian prisons and airports encountered a new force in the campaign to protect public morals – a team of highly trained gerbils. The sensitive noses of these furry detectives won £20,000 worth of government investment to help track down the nation's criminals.

A number of problems inherent in gathering a workforce of conventional dope-sniffers (dogs, for example) led the Canadians to make the breakthrough for rodent liberation. Dogs obey only one handler (or two at most); they eat a lot; and they need a fair amount of space and care. Add to that the discomfort they bring by sniffing around the ankles of harassed travellers in airports and there seemed to be a good case for the gerbil as a seeker-out of contraband.

The customs authorities set up an intensive training scheme to make the most of the gerbils' talents. At airports, where a few pioneering gerbils were in action, the sniffers crouched in their cages behind a counter and caught the smells of travellers as they walked past a fan. Although the customs gerbils faced retirement sooner than a dog would have done, officials were enthusiastic about their cheaper upkeep and their modest approach to industrial relations.

⚛ Wasps – sniffer dogs with wings?

In 2006, after three 10-second training sessions, Glen Rains'
crack team of sniffers was ready for anything. They could be
co-opted into the hunt for a corpse. They might join the search
for a stash of Semtex or a consignment of drugs. Or they
could have the more tedious job of checking luggage at the
airport. Whatever the assignment, their role was the same: to
pick up a scent no human nose could detect and pinpoint its
source. These recruits to the fight against crime were smaller,
cheaper and more versatile than a sniffer dog, and more sensitive than an electronic 'nose'. They were wasps.

Insects have exquisitely sensitive olfactory systems. Their
antennae are covered with microscopic sensors that can
detect the faintest odour. Some are also remarkably quick
learners. So it was hardly surprising that they aroused the
interest of the military and security services, police and
customs, all badly in need of ultra-sensitive, flexible and portable odour detectors. Insects obviously have the right stuff,
but could they use it to sniff out smells they would never
encounter in nature – a hint of explosives, say, or a whiff of
cocaine? And if so, would it be possible to make a practical
device that harnessed their skills?

Enter Wasp Hound, a hand-held odour detector with a
team of little black wasps as its sensor. Developed by Rains, a
biological engineer at the University of Georgia, his colleague
Sam Utley and Joe Lewis, an entomologist at the US Department of Agriculture's Agricultural Research Service in Tifton,
Georgia, the device was only a prototype, but the team had
high hopes for it.

Did this mean the days of the sniffer dog were numbered?
Probably not. 'We don't see insects as a replacement for dogs,'
said Rains. 'But they do have lots of advantages. They cost
pennies to raise. They don't need special handling, and

because they are so quick to train you can have them on call, ready to learn a new smell whenever you need it.'

Most of us don't really like wasps and consider them best avoided. Still, you'd think most of us wouldn't want sheep foetus injected into our bottoms. Yet …

Germany bans 'rejuvenating' sheep cell injections

In 1997, Germans were no longer permitted to have tissue from sheep foetuses injected into their buttocks. Several thousand people in Germany underwent this process every year, believing that the foetal cells had rejuvenating properties. But in the late 1990s, health officials in Bonn said that they had decided to ban the practice, pointing to severe immune reactions in some patients. A Ministry of Health report said that up to 5 per cent of patients had reactions to the injections and in five documented cases patients had died.

Jutta Buscha, a doctor in the Bavarian town of Rottach-Egern who practised the therapy, said 'I don't understand this enmity, this persecution. The people who sit in judgment on us never even came for a visit.'

And finally to Zoe. Zoe, too, could be said to be a bit of a pain in the bum. And he, it has to be noted, is not really an example of mad research. But Zoe did become very popular among New Scientist's *readership when his intriguing plight first appeared in the magazine. And we reckon his story deserves another airing …*

❊ Humming dog

You would always know when Zoe the West Highland terrier was around. Zoe had two problems in life. One was that, despite the name, Zoe was male. The other was that a persistent humming noise emanated from his head – a humming noise that was audible to Zoe's owners and any other humans who happen to be around.

In 1996, his mystified owners took Zoe to a vet, Ian Millar of Belfast. After various tests and a course of treatment with antibiotics in case an infection was responsible, Millar professed himself mystified as well. He wrote to *The Veterinary Record* asking if anyone could shed light on the problem.

A couple of weeks later, Patrick Burke of the University of Edinburgh's Royal School of Veterinary Studies wrote back. The problem, he suggested, was a phenomenon known as 'otoacoustic emission'. In this condition, the normal hearing pathways in the ears are somehow reversed, so that the cochlear efferent nerve fibres stimulate outer hair cells to vibrate and make a noise. Other parts of the ear, such as the tympanic membrane, can then amplify the sound, until you end up like Zoe, humming wherever you go.

8 The world looked different back then

When novelist L. P. Hartley wrote 'The past is a foreign country, they do things differently there', it's almost certain he didn't have the field of scientific endeavour in mind. But boy, did it look different back then ... We believed there was nothing we wouldn't be able to do in the future. Pesticides would fix everything. As would nuclear power. And then there were atomic bombs. Lots of atomic bombs.

When William Gladstone, then Chancellor of the Exchequer, was invited to a demonstration of Michael Faraday's equipment for generating the new scientific wonder of electricity in the second half of the 19th century, he watched carefully, stood silent for a moment and then asked Faraday: 'It's very interesting, but what practical worth has it?' 'One day, sir,' replied Faraday, 'you may tax it'.

Faraday, being the clever sort he was – and realising that, as Benjamin Franklin's oft-quoted saying goes, 'Nothing can be said to be certain, except death and taxes' – correctly predicted the future. His fellow researchers who appear in this chapter were less prescient.

We now know how deadly pesticides can be. We now know that a nuclear reactor in a bucket might not be so clever after all. And as for the paperless office which has been predicted many times over – next century anyone? We can guarantee that whatever seems a perfectly reasonable prophecy today will only turn out to be the subject of this chapter's descendent in 50 years' time. *New Scientist* has reported on so many firsts down the years, all which seem to offer the possibility of a golden future, yet most of which end up being more like base lead. Remember that for every Kyoto protocol there's a George W. Bush.

Back when *New Scientist* was launched, science and technology were seen as unalloyed forces for good. As we've pointed out, this optimism applied especially to nuclear power: soon we would fly in atomic aircraft, wear nuclear-powered watches and clean clothes by spraying them with radiation. Three-Mile Island and Chernobyl put us straight on that one, to say nothing of the likes of feeding cows the remains of other cattle, leading to the BSE crisis and new variant Creutzfeldt-Jakob disease in humans. All of which have dented the public's enthusiasm for science's prophecies of jam tomorrow.

Of course, what follows reflects the sensibilities of *New Scientist* and the limited predictive powers of its journalists down the decades as much as it does the naivety of the scientists it reported on. But, of course, we are all products of our times.

But perhaps more than anything this chapter is a warning to all those who want to live beyond the here and now. Forget it. Just keep filling in your tax return and put a sum aside each month for funeral costs and death duty. That way, as Ben Franklin wisely predicted, you won't end up looking silly …

Sometimes, the world looked so different that New Scientist *was frequently a highly sceptical beast. Among many things we said couldn't or wouldn't happen, like colour television, commuting by aircraft and that damned paperless office (and we are still right on that one) was this 1957 editorial on space travel.*

🌐 How soon to the Moon?

Though the launching of the first small satellite – *Sputnik 1* – into Earth orbit is a necessary first step towards interplanetary travel, it is but a small step towards the solution of a much greater problem, one that bristles with innumerable difficulties and complexities. Though within a few years we may see the launching of a small vehicle that will either impact the Moon or will circle it and return to Earth, it is very likely that generations will pass before man ever lands on the Moon and that, should he succeed in doing so, there would be little hope of his returning to Earth and telling us of his experiences. Beyond the Moon, almost certainly, he is never likely to go.

(10 October 1957)

Twelve years later humans had walked on the Moon. Still, we shouldn't beat ourselves up about our dreadful powers of prediction. That great rocket scientist and former employee of the Third Reich Wernher von Braun was also wide of the mark, although at least he was over-optimistic, unlike New Scientist.

🌐 Space exploration in 1984

Lunar landings will have long since passed from fantastic achievement to routine occurrence. Astronauts will be shuttling back and forth on regular schedules from the Earth to a small permanent base of operations on the Moon. A part of the activity on the lunar surface may well be the operation of an astronomical observatory, taking advantage of the favourable observation conditions there. Private industry will have entered Earth-orbital operations on a large scale.

The existence of a low order of life on Mars will probably have been proven, and the significance of the seasonal changes of the Martian canals established.

(16 April 1964)

We were, however, terribly, optimistically wide of the mark about the uses for nuclear power and its offspring, the atom bomb. And even the notorious atomic accident at Windscale didn't put us off.

A hazard to health?

The accident at Windscale has aroused questions about the safety of atomic power stations. We believe atomic industry to be no more dangerous than conventional. Radioactivity is easily detected, some chemical hazards are not.

(24 October 1957)

Atom bombs to release oil?

American atomic physicists are giving much thought to peaceful uses for nuclear bombs. Of the several suggestions that have been publicly mentioned, those concerned with the release of underground oil deposits are most common. One such scheme has just been announced by the United States Bureau of Mines.

Although precise details of this and similar plans have not been disclosed, enough is known of the physical effects of underground nuclear explosions to permit reasonable forecasts of what will happen when a bomb is detonated in various types of oil deposit.

(15 January 1959)

Bonkers? They didn't think so way back when. Atom bombs were expected to have great societal benefits. You could even dig the Panama Canal with them. Couldn't you?

⚛ Nuclear digging on trial

Project Gasbuggy, the world's first commercially sponsored nuclear explosion, is scheduled to take place on 14 November 1967 on a lonely plateau in northern New Mexico. Gasbuggy is the most advanced test yet in the US Atomic Energy Commission's Plowshare programme, which aims to find peaceful uses for nuclear explosions. Its object is to determine how effectively nuclear explosions can release natural gas from normally impermeable rock. Geologists estimate that successful use of the technique could double the usable gas reserves of the United States.

In the test, a 26-kiloton nuclear explosive will be detonated 4240 feet below the surface, just under a layer of gas-bearing rock 300 feet thick. The blast is expected to create a 'chimney' of broken rock, greatly increasing the flow of natural gas through a well to the surface.

The most troublesome problem is radioactivity – not from the venting of the explosion, but from the isotopes that will be created in the underground chimney. When the explosive goes off, it will produce a bubble 160 feet in diameter. As the cavity cools, its ceiling will collapse to create the chimney. Most of the radioisotopes produced by the blast will disappear quickly, but three will remain – iodine-131, krypton-85 and tritium. The iodine will decay to a stable form within months and the krypton may be trapped in the molten rock that flows to the bottom of the chimney, but the tritium will linger on.

Two tests similar to Gasbuggy are scheduled for next year in formations different from those in New Mexico. The AEC is even moving ahead, rather gingerly, with the controversial programme of nuclear excavation. Two tests are designed to gather data on the possible nuclear excavation of a new Panama Canal. Since they would release radioactivity on the surface, the tests were postponed to avoid upsetting US-Soviet talks on a treaty to prevent the spread of nuclear weapons.

There are also proposals for using nuclear explosions to release oil from impermeable formations and for nuclear exploitation of shale oil deposits. More radical ideas are waiting on the results of these experiments. The arid state of Arizona, for instance, is studying the possibility of using nuclear explosives to trap rainfall that now evaporates. The explosives could, it is thought, open tunnels for water to run off into underground aquifers.

(26 October 1967)

The Gasbuggy bomb was eventually detonated on 10 December 1967. It did stimulate greater gas flow, but uncertainty remained over the size of the improvement. Public opposition grew – nobody would buy gas that might be contaminated with radioactivity, for example – and in June 1975, after 35 nuclear detonations, the Plowshare programme was wound up. How naive we were. We even thought waste radioactive products sounded useful.

⚛ Atom ash keeps cloth clean

One of the substances present in the radioactive ash from atomic power stations, strontium-90, is now being used by manufacturers of woven and knitted fabrics to overcome fog markings. During the winter, dirty lines are likely to appear on the fabric every time a machine is stopped for lunch breaks, or at the end of the day's run. The dirt gathers because the cloth becomes charged with static electricity and attracts particles of dirt from the atmosphere.

Various methods have been used to prevent the fabric becoming charged with electricity. The most widely known is to use an anti-static dressing (based possibly on a diamine fatty alkyl sulphate dissolved in a mixture of mineral and vegetable oils). Another idea is to use a radioactive element.

This element, a simple bar of metal containing the radio-active substance, discharges the electricity on the fabric by spraying it with atomic particles. What happens, in effect, is that the air around the fabric is made conductive so that the electric charge is carried away.

The original development of units of this sort was made by the Shirley Institute, using thallium-204 as the radioactive substance. The new elements containing strontium-90 have many advantages: the strontium decays less rapidly than thallium, and the radiation emitted has a rather higher energy, so it can penetrate denser fabrics.

(6 December 1956)

Back then we didn't seem to worry at all about how dangerous strontium-90 might be. We were even happy to wear it.

Nuclear battery for electric watches

An American company is now offering a nuclear battery for sale. It is similar to one developed a few years ago in the US, making use of beta rays from a radioactive source to bombard a semi-conductor, which produces a steady but small electric current.

The new nuclear battery, made by Walter Kidde Laboratories in conjunction with the Elgin National Watch Company, uses promethium-147, a by-product of nuclear fission, as its source of beta rays. An earlier battery developed by the Bell Telephone Company used a strontium-90 source and developed sufficient power to run miniature electronic equipment. The Walter Kidde battery is designed for use in electronic watches and certain electronic applications.

(28 March 1957)

✳️ Nuclear reactor in a bucket

A nuclear reactor small enough to fit in a household bucket has been developed by the German firms of Siemens, BBC and Interatom. It is known as the Incore Thermionic Reactor and is intended for future space vehicles. The reactor uses highly enriched uranium as fissile material and liquid sodium as coolant. In contrast to most other nuclear methods of producing electrical power, there are no intermediate moving parts – the current is produced directly by a thermionic effect.

During fission the interior reaches 1400 °C and thermionic electrons are ejected from the tungsten which coats the fuel. These jump across the gap to the surrounding cylinder and produce an electric current. The total weight is less than 1000 kilograms.

Such power supplies could be of great importance in powering space vehicles as, unlike solar cells, their output is not subject to an upper limit based on surface area. It may be possible to use them to provide the power to navigate a vehicle into a stationary orbit and even to transmit television signals with enough power for them to be received directly by a viewer on Earth.

Further development is expected to take from four to six years and a test installation is to be built at the nuclear centre at Jülich.

(9 May 1968)

⚛ Atomic aircraft

Enthusiasm for the nuclear-powered bomber project in the United States blows alternately hot and cold. Mr R. E. Gross, chairman of Lockheed Aircraft, one of the two companies with contracts to develop the airframes (the other being Convair), has said recently that if the American government were to give the 'go-ahead signal', Lockheed could have an aircraft ready to make its first flight in the mid-1960s.

The type of aircraft the company has in mind would have the shielded crew cabin in the nose, the reactor in the tail as far from the crew as possible, a small tankage of conventional turbine fuel for take-off and landing so that the reactor was only at full power in the air and never near the ground, and thin straight wings free from the encumbrances of fuel tanks, engines or undercarriage gear. The US Air Force wants atomic bombers of this kind for the same reason that the navy wanted atomic submarines: they could range the world without refuelling.

But the air force faces one great technical difficulty that did not trouble the navy – weight. Even when the weight of reactor shielding is cut to the minimum by concentrating on a radiation-proof cabin for the crew rather than trying to block all escape of radiation from the reactor, it still remains the biggest barrier to getting an atomic aircraft off the ground.

And in spite of the unlimited range that only a nuclear plant can give, some scientists believe it is not a development that should be undertaken at this stage. Mr Cleveland, who is in charge of Lockheed's atomic design, has himself suggested there are serious health problems connected with the maintenance of atomic aircraft because of the radiation leakage. Other experts have pointed to the hazard that would follow the crash of an atomic aircraft, whose reactor would almost inevitably be cracked open, making rescue all but impossible and, if there were a fire, spreading fission products downwind from the wreckage.

(11 July 1957)

Fortunately, the project never ... er ... took off. But what about trains?

⚛ Trains to go nuclear

Plans are being prepared by the German railways for an atomic locomotive powered by a gas-cooled reactor using enriched uranium.

The plans envisage a locomotive 35 metres long and 3 metres wide, which will weigh approximately 185 tonnes with an output of 5916 horsepower. The vehicle will be supported on eight axles. The gas-cooled reactor will make the locomotive much lighter than previously suggested designs because it will not require a refrigerator car and it will be able to omit a number of secondary safety devices – 50 tonnes are saved by using helium as a cooling agent for the reactor and a further 90 tonnes are saved by a reduction in the secondary safety devices. A further weight saving is also possible if the designers elect to use mechanical-hydraulic transmission instead of the heavier electrical type. The reactor will be a Babcock and Wilcox design that is 305 millimetres long.

Instruments will be installed in the locomotive to detect and measure radioactive emanation from the power plant. The crew will be provided with special clothing, and the driving compartment will be insulated against radiation, noise and heat.

The running costs of the locomotive are expected to be lower than the cost of operating a steam locomotive in Western Germany but somewhat higher than an electric one.

No information is yet available about when construction will begin nor when the new design is likely to come into passenger service.

(24 January 1957)

Still, even if we did have fears about radiation – which it seems we did not – it was hoped they could be eased by turning us into Batman.

✹ Bat blasting

When the Argonne National Laboratory – one of the four major research establishments of the US Atomic Energy Commission – announced five years ago that bats had survived huge doses of radiation there was considerable excitement. For, it was reasoned, if the source of the bat's abnormally high resistance to radiation could be established, it might point the way to increasing the resistance of other mammals, including humans.

(10 January 1957)

It turned out that only hibernating bats survived high radiation. Once they woke up, it wasn't so pretty. But our unbridled optimism that we could make radiation-resistant humans can be construed as either terribly sweet or terribly worrisome. And we are quite certain the bats didn't enjoy finding out. Eventually, though, we began to cotton on to the fact that radiation perhaps wasn't the panacea for the world's ills.

⚛ Radiation – a losing game?

'It is apparent that the atomic dice are loaded. The percentages are against us and we ought not to play unless we must'. This was the sharply worded warning on the lead page of the American journal *Science* this week.

It was a comment on a research report submitted by Dr E. B. Lewis of the California Institute of Technology in Pasadena. What the young biologist had announced was something that scientists had been trying to discover since the first atom bomb exploded in 1945. There is a direct linear relationship between the amount of radiation received by a person and the occurrence of leukaemia, a fatal disease of the white blood cells. Lewis demonstrated that if the general population were to ingest the amount of strontium-90 which the radiation effects panel of the National Academy of Sciences last June pronounced to be a safe maximum dose, somewhere between 150 and 3,000 more people would die of leukaemia each year in the United States alone.

'Thanks to Lewis,' wrote the editorial, 'it is now possible to calculate within narrow limits how many deaths from leukaemia will result in any population from fall-out or other source of radiation. We are approaching the point at which it will be possible to make the phrase "calculated risk" for radiation mean something a good deal more precise than "best guess".'

(16 May 1957)

Cue the sound of walls crumbling around the atomic research community. But we weren't always behind the game. There were a few innovations we trumpeted with wide-eyed amazement which we now encounter day-in, day-out. How soon the miracles of yesterday become the commonplaces of today. Well, they seemed eccentric back then …

✳ Jumping the lights

De-rationing of petrol has brought back the traffic jam to our cities and towns. All the motor manufacturers are producing to the limit of their capacity and, even if as many as half their cars are destined for export, the remainder will be more than sufficient before long to cause complete congestion in some places.

Already in the last thirty years automatic traffic signals have revolutionised traffic control. First installed in Leeds (where they are still referred to as 't robots'), they were originally designed to work on a fixed time cycle – so long allowed to one traffic stream and so long to the other. More recently signals have been developed which are operated partly by the traffic itself, which indicates its presence by running over a rubber pad.

(23 May 1957)

✳ Sorry, we can't take your call ...

The telephone in the doctor's surgery, a taxi service, or even the grocery will always be answered – even when no one is in – now that the ansaphone has received Post Office sanction and becomes generally available in Britain. The machine, roughly the size of a portable gramophone, was developed by Southern Instruments (Communications) Ltd and, in common with most telephone equipment, is to be installed on a rental basis. The minimum cost of rental contracts including service and maintenance is £1 2s 0d a week.

(22 January 1959)

⚛ Video, video

Remarkable though the 'instant' processes for black-and-white and colour photography appear today, they are by no means the last word in techniques for the amateur. Simultaneous recording of sight and sound on tape is just one of the coming methods we expect to see.

(8 April 1965)

Back in 1967 New Scientist *had a rather outdated attitude to gender roles.*

⚛ Shopping in disguise

The all-purpose credit card usable in stores, hotels and so on has been a feature of the American scene for some time and grows in popularity in Britain. In the United States, however, they intend to go one better. The customer at, say, a supermarket will have to pop the credit card into an electronic device which will signal a computer centre to check on his or her bank account. Not so far out, you think? Well it gets better. The computer will also give the salesgirl a visual description of the holder of the card: colour of hair and eyes, height and weight, and so on.

If the description tallies, the customer gets the goods. However, I foresee problems. Most shopping is still done by women – so how will the computer know that this season Madam is sporting a blonde wig or a blue rinse, or wearing skyscraper heels, or weight-disguising falsies, or has changed the colour of her eyes with contact lenses?

(21 December 1967)

⚛ In the tube

The idea of 'test-tube babies' is no longer something to be woven into the plot of a science fiction novel. Serious-minded scientists are not only thinking about cultivating human embryos on the laboratory bench – they are developing the techniques which will make this a practical possibility. In the current issue of *The Lancet* Dr R. G. Edwards, now at the Physiological Laboratory in the University of Cambridge, describes work done there and at Johns Hopkins Hospital, Baltimore, US. Dr Edwards has taken eggs from ovaries, or parts of ovaries, obtained from 16 women who for one reason or another had to have these organs removed.

(11 November 1965)

And, just as interestingly, the battles of yesteryear are still being fought out today. Here we offer a cynical take on what was at the time – and arguably still is – the most sensitive area of scientific research …

⚛ Bad breeding

The Vatican's official press spokesman yesterday denounced as 'immoral acts and absolutely illicit' experiments in Britain in fertilising human eggs outside the body.

Does the current turmoil over test-tube babies, which has spilled across so many acres of newsprint, and occupied so many broadcasting hours, mean that the public has at last begun to take science seriously? Almost certainly not. What has happened is that editors have realised that apparent 'breakthroughs' in medicine hold a fascination for the reading, listening and viewing masses which at least equals

that exerted by tales of sex murders, and other titillating aber-
rations of human behaviour.

(20 February 1969)

*But enough conceit about our successful visions of the future.
What about when we got it all very wrong?*

☀ TV hell

Telephone, radio, and television systems are capable of con-
veying information at a far greater rate than men and women
can cope with it. This is shown by some recent measurements
involving an application of 'information theory' to human
beings. 'Information' in this technical sense includes all the
entertainment and small change of ordinary conversation.

(23 May 1957)

☀ No chance

What is the future for colour television? This is one of the
questions which must perplex the Pilkington Committee on
Broadcasting, and which – for that matter – must keep Mr
Bevins, the Postmaster-General, tossing uneasily in his sleep.
For as the days get shorter and as those persons known as
'viewers' begin to recognise that the content of their patch of
fluorescent flicker is much the same as it was a year ago, the
clamour for colour television must inevitably increase.
Already some papers have called Mr Bevins 'colour-blind'. It
will be surprising if he is not much more colourfully described
before the winter is over. The cry will be for a nationwide
broadcasting service supplying a coloured television picture
to virtually every house in the country.

This should be allowed to echo unheeded in the ether. For it seems that nobody has so far been able to demonstrate that a colour television service will benefit anybody but the television maintenance men.

(16 November 1961)

We were still sceptical eight years later. Heaven knows what we would have made of HD TVs.

The last word on colour television

Even after they've surmounted the paramount problem of raising enough money to pay for the latest status symbol, owners of colour television sets are going to be in for a few headaches living with them. If you haven't got a disused photographic darkroom on your premises, then you'll have to put your front room into a state of penumbra fit for a fraudulent séance or World War III if you want to have colours coming up good. The set must be degaussed and aligned carefully in the Earth's magnetic field. Stray magnetic fields can play havoc with your tinted picture, and so large, ferrous objects such as comforting radiators and knights-in-armour with fire-irons in their bellies must be removed from the sacred vicinity, and all electric clocks, vacuum cleaners and pocket magnets declared persona non grata. And when the whole monster has been professionally set up, it's going to be at least twice as difficult for the twiddling owner to tune as was the black-and-white steam model.

(7 March 1968)

So, weirdly sceptical about colour TVs, but wildly optimistic about any number of other mod-cons. We still couldn't get it right.

✸ Masterchef

In the household of 1977 there will be a prepared food unit. Frozen food packages will be stored in classified ranks below four ovens. In the morning you will set the time you want to eat, the meal containing up to four food items. At the correct time the unit will transfer the four food packages from freezer to oven. They will be ready simultaneously at your chosen mealtime.

(21 November 1957)

✸ Tumble fryer

After 20 years of research, Paul Groves of Gloucester claims to have made a kitchen appliance that cooks, washes up and even washes clothes.

On the outside it looks like a chest freezer, but inside is a heating chamber and a revolving drum. Supplies of cooking oil, water and detergent are held in tanks. You put the food in the drum, and as it turns, vanes draw in hot air from the heating chamber to cook it. During cooking, food can be held stationary or tumbled in the drum. For frying, oil is injected as a mist. For boiling, a water heater produces steam.

When the meal is over, you put the dishes back in the same drum and switch the device to pump in water and detergent. A similar cycle is used to wash clothes. Groves says he expects the machine to cost about £800.

(26 August 1995)

We expect it to vanish. But then we would, because we know what happened. Outside the home, we thought we'd be eschewing the bus, and heading for the office at high speed …

✵ What? No delays?

Any calculation of the future traffic requirements in 'commuter country' such as those parts of England south of the Thames must take into account the part which may be played by vertical take-off aircraft. Air Chief Marshal Sir Ralph Cochrane has now calculated the volume of traffic likely to be travelling at peak hours in relation to running commuters' aircraft at 600 miles per hour between 200 take-off pads at distances of 60 to 120 miles from London and a similar number at terminals near the main railway stations in the capital. He has calculated that in 15 years the population of this region of England will have overloaded the capacity of the railway at rush hour. His proposal is for vertical take-off aircraft to meet the increased demand using modern guidance systems that can keep an aircraft on track 'with an error measure in yards'.

Aircraft to London would depart when the London terminal was clear and within 14 minutes they would unload 110 passengers at an allotted pad, then refuel and be turned around within 7 minutes – 60,000 passengers could be handled in an hour. Terminals would be no bigger than an acre and 360 aircraft would be required.

Instead of sitting on the stopping train the daily commute would be at 10 miles per minute by the end of the 20th century, a legitimate set of expectations which can no longer be ignored when envisaging the way of life in the future.

(16 February 1961)

In those days we knew how to fix the world's problems …

✳ Locust plagues ended

The world may have seen the last of the ancient scourge of the locust. This claim is being made by experts at the United Nations Food and Agriculture Organisation in Rome. They believe that international satellites that photograph the surface of the Earth mean that any sudden infestation of locusts can be nipped in the bud using modern pesticides.

Satellites can even spot conditions where the desert locusts are likely to breed rapidly and move into crop-growing areas, where a single swarm can consume 80,000 tonnes of corn a day – enough to feed 400,000 people for a whole year.

(21 April 1983)

… and if we didn't know how to fix earthbound horrors, we could always move somewhere else. Somewhere the grass was going to be greener.

✳ Little green plants

William M. Sinton of the Smithsonian Astrophysical Observatory thinks it 'extremely probable' that vegetation of some form is present on Mars. He describes in the *Astrophysical Journal* how he studied the infrared light reflected from Mars when it made its close approach to the Earth last year.

He found a significant dip in the spectrum – a weakening in the reflection at a wavelength of 3.46 thousandths of a millimetre. This is just what one would expect if large molecules containing carbon and hydrogen are present.

Large organic molecules do not necessarily imply life. It seems unlikely, however, that organic molecules would remain on the Martian surface without being covered by dust from storms or being decomposed by the action of the solar ultraviolet, unless they possessed some regenerative power.

Here then, is the confirmation that the dark patches on Mars which wax and wane on its arid surface with the passing seasons consist of some form of plant life.

(28 November 1957)

Oh dear. But it's all so easy when we know better isn't it? The naivety of the past only looks like naivety today because we are a bunch of clever clogs with hindsight. Even so, you would have thought that, back in 1957, people would have known it wasn't a good idea to scoff a mushroom you'd found up a Mexican mountain, although we're glad somebody did. Roger Heim is a real oddball hero.

⚛ Divine mushrooms

The French Academy of Science was told last week by Professor Roger Heim, Director of the Paris Natural History Museum, of the extraordinary effects of a Mexican mushroom called 'teonanacatl'. He found the fungus on a mountainside in southern Mexico.

The species known as 'divine mushrooms' was discovered by Mr Gordon Wasson, an American amateur mycologist, who has read of vision-producing fungi in old Spanish documents. He devoted ten years to trying to chase them. When eventually he succeeded in finding them and confirmed their strange vision-producing properties, he called in Professor Heim, one of the world's leading mycologists, to investigate and identify the fungus.

In Professor Heim's own words this is what happened when he ate a spoonful of the fungus. 'Only a few seconds after I had swallowed my portion,' he told the Academy, 'I became subject to the optical and physical anomalies. The colours of the room in which I was became brighter. I had double vision. I suddenly felt in an extremely mirthful mood. During more than two hours a sort of fantasy in blue unfolded before my eyes.'

Back in Paris, Professor Heim twice repeated the experiment with the same results. The visions, he explained, were caused by a drug in the fungus, probably of the opiate variety. He thinks that an overdose of 'divine mushrooms' would cause mental derangement.

(15 August 1957)

... if the recipient wasn't deranged already. And our own naivety showed through in 1959. It seems we'd been concerned about the effects of TV, especially on our younger generation. For once, we came out in favour of the gogglebox ...

✵ TV is our friend

Four years ago, when television was still a myth to most people and an intriguing novelty to a few million BBC viewers, it was quite widely feared as being likely to breed a crop of social ills. Parents were particularly bothered about the impact of television on their children.

Today we can feel reassured somewhat. At the suggestion of the BBC's Audience Research Department, the Nuffield Foundation 'decided to investigate whether by use of scientific methods it was possible to elicit some objective facts that would confirm or deny prevailing impressions', and the results of this eliciting operation have now been published in a vast report at two guineas. It seems that the influence of television is considerably less dramatic than was feared, that most children absorb it without strain, except those who are of low intelligence or who are already afflicted with emotional problems or with disturbed families.

(22 January 1959)

Today's parents will no doubt be delighted to know that it's OK to let their kids watch TV – all they need to do is have their IQ tested or their child psychologically analysed first. So, finally, on to our treatment of women in decades past. We've already touched on paternalistic attitudes earlier in this chapter, and now it's time to come clean. The world really was a different place all those years ago and New Scientist was as hideously sexist as they come, even when we were trying desperately to be right-on ...

✳ The under-employed sex

So far as grey matter is concerned, both males and females start out with equal chances. When the educationalists apply their IQ tests, no distinct difference between the sexes emerges. A look at any class in a mixed grammar school makes this obvious. All have passed their 11-plus, and are therefore supposed to be in the upper intelligence bracket. Half the class will be male; half female. Were it overloaded with males, there might be some justification for industry to operate a system of male supremacy.

An individual with Applied Systems and Personnel told me 'girls sometimes make better programmers than boys because they have more patience and are more meticulous. An intelligent girl who has the patience to do embroidery has just the right mentality to do the job'.

(29 February 1968)

And back in those days, you really could print ads like the ones that follow without being hauled off for gender-attitude realignment therapy. These are so out-there that they've fallen over the edge of their flat Earth.

✳ Careers in marine geology

There are vacancies for three geologists (one senior and two junior) with a sedimentological bent, and an affinity for the sea. Experience in sailing, surfing, skin diving or seafaring is an advantage but not a necessity, but adaptability and alertness are essential. There will be a pre-fab house for the senior man if married. For the junior posts single men are preferred, but married men are not debarred providing the wives can adapt themselves to rugged semi-camping conditions. The pioneering spirit is as important in the wife as in the husband. An interest in nature generally is a great asset to either. It is hoped to interview husband and wife together.

(9 April 1964)

✳ Chemists are required

Graduate chemists with good honours degrees are required by this Group of Companies for both organic and inorganic positions, in the London and Midland areas. Commencing salaries attractive and progressive. Non-contributory pension scheme, assistance given with house purchase. A female would be considered.

(24 August 1961)

Wonder if she'd be able to run a paperless office? Ah yes, the paperless office. We are sure we predicted it more than once, we just can't seem to locate the hard-copy evidence under the pile of old magazines and memos lying on the desk …

Index

DEC 02 2010

DEC 28 2010

BASEMENT

inbox (you have new knowledge)